INSTITUTE OF GEOLOGICAL SCIENCES
Natural Environment Research Council

Mineral Assessment Report 80

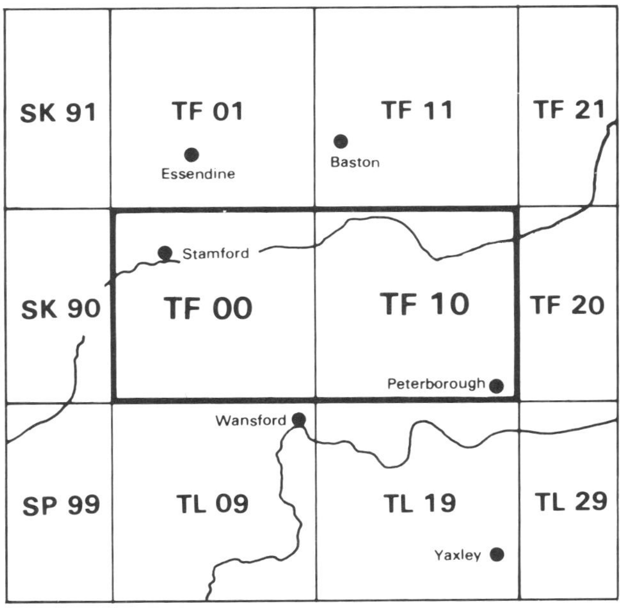

The sand and gravel resources of the country between Stamford and Peterborough

Description of 1:25 000 resource sheets TF 00 and TF 10

S. J. Booth

Contributors
R. J. Wyatt and J. B. L. Wild

ISBN 0 11 884180 7*

London Her Majesty's Stationery Office 1981

The first twelve reports on the assessment of British sand and gravel resources appeared in the Report series of the Institute of Geological Sciences as a subseries. Report 13 and subsequent reports appear as Mineral Assessment Reports of the Institute.

Details of published reports appear at the end of this report.

Any enquiries concerning this report may be addressed to Head, Industrial Minerals Assessment Unit, Institute of Geological Sciences, Keyworth, Nottingham NG12 5GG.

PREFACE

National resources of many industrial minerals may seem so large that stocktaking appears unnecessary, but the demand for minerals and for land for all purposes is intensifying and it has become increasingly clear in recent years that regional assessments of the resources of these minerals should be undertaken. The publication of information about the quantity and quality of deposits over large areas is intended to provide a comprehensive factual background against which planning decisions can be made.

Sand and gravel, considered together as naturally occurring aggregate, was selected as the bulk mineral demanding the most urgent attention, initially in the south-east of England, where about half the national output is won and very few sources of alternative aggregates are available. Following a short feasibility project, initiated in 1966 by the Ministry of Land and Natural Resources, the Industrial Minerals Assessment Unit (formerly the Mineral Assessment Unit) began systematic surveys in 1968. The work is now being financed by the Department of the Environment and is being undertaken with the cooperation of the Sand and Gravel Association of Great Britain.

This report describes the resources of sand and gravel of 200 km^2 of country between Stamford, Lincolnshire, and Peterborough, Cambridgeshire, shown on the accompanying composite 1:25 000 resource map TF 00 and TF 10. The survey was conducted during 1976–1977 by S. J. Booth who supervised the drilling and sampling programme, assisted by J. L. Knight and E. R. Moczarski. S. J. Booth compiled the report assisted by J. B. L. Wild who also contributed to the section on the composition of the sand and gravel.

The work is based on six-inch geological surveys by members of the Institute's Field Staff. Ground east of grid Easting 12 was surveyed between 1939 and 1947 and published as part of the Stamford One-Inch New Series Geological Sheet 157, and the area of Peterborough New Town was surveyed in 1968 and published on the 1:25 000 scale with an accompanying geological description in 1974. A six-inch survey of the remaining areas was specifically commissioned by the Department of the Environment and completed in 1977 by R. J. Wyatt who also re-interpreted the 'Fen Gravel(s)' and provided a contribution which formed the basis of the geology section.

Officers of the Property Services Agency based at Newmarket were responsible on behalf of the Institute for negotiating access to land for drilling. The ready cooperation of landowners, tenants and gravel companies in this work and the assistance of officials of the Anglian Water Authority, East Midlands Electricity Board, East Midlands Gas Council, Peterborough Development Corporation and the Cambridgeshire and Lincolnshire County Councils are gratefully acknowledged.

G. M. Brown
Director

Institute of Geological Sciences
Exhibition Road
London SW7 2DE

15 December 1980

CONTENTS

FIGURES

MAP

TABLES

The sand and gravel resources of the country between Stamford, Lincolnshire, and Peterborough, Cambridgeshire

Description of 1:25 000 sheets TF 00 and TF 10

S. J. Booth (with contributions by R. J. Wyatt and J. B. L. Wild)

SUMMARY

The geological maps of the Institute of Geological Sciences, pre-existing borehole information, and 101 boreholes drilled for the Industrial Minerals Assessment Unit, form the basis of the assessment of sand and gravel resources in the country between Stamford, Lincolnshire, and Peterborough, Cambridgeshire.

All deposits in the area which might be potentially workable for sand and gravel have been investigated and a simple statistical method has been used to estimate the volume.

The accompanying 1:25 000 map is divided into five main resource blocks containing between 1.7 and 31.0 km^2 of sand and gravel. The geology of the deposits is described within each block and the mineral-bearing areas are distinguished by sub-blocks. The mean thicknesses of overburden and mineral and the mean gradings, together with detailed borehole data are also given. The geological lines and symbols, the positions of boreholes used in the assessment, grading information and the outlines of the resource blocks and sub-blocks are shown on the accompanying map.

Bibliographical reference

Booth, S. J. 1981. The sand and gravel resources of the country between Stamford, Lincolnshire, and Peterborough, Cambridgeshire. Description of 1:25 000 sheets TF 00 and TF 10. *Miner. Assess. Rep. Inst. Geol. Sci.*, No. 80.

Author and contributors

S. J. Booth, BSc
J. B. L. Wild, BSc
Institute of Geological Sciences,
Keyworth, Nottingham NG12 5GG

R. J. Wyatt
Institute of Geological Sciences,
Exhibition Road, London SW7 2DE

Note

National Grid references are given in square brackets. In this publication all lie within the 100-km square TF.

INTRODUCTION

The survey is concerned with the estimation of resources, which include deposits that are not currently exploitable but have a foreseeable use, rather than reserves, which can only be assessed in the light of current, locally prevailing, economic considerations. Clearly, both the economic and social factors used to decide whether a deposit may be workable in the future cannot be predicted; they are likely to change with time. Deposits not currently economically workable may be exploited as demand increases, as higher grade or alternative materials become scarce, or as improved processing techniques are applied to them. The improved knowledge of the main physical properties of the resource and their variability which this survey seeks to provide, will add significantly to the factual background against which planning policies can be decided (Archer, 1969; Thurrell, 1971; Harris and others, 1974).

In this report the assessment is in most cases presented at the 'indicated' level of assurance. However, in those areas where the available information is not sufficient, the assessment is conducted at the 'inferred' level (see also Appendix B, paragraph 12). In the former "tonnage and grade are computed partly from specific measurements, samples, or production data and partly from projection for a reasonable distance on geologic evidence. The sites available for inspection, measurement, and sampling are too widely or otherwise inappropriately spaced to permit the mineral bodies to be outlined completely or the grade established throughout".

At the 'inferred' level "quantitative estimates are based largely on broad knowledge of the geologic character of the deposit and for which there are few, if any, samples or measurements. The estimates are based on an assumed continuity or repetition, of which there is geologic evidence: this evidence may include comparison with deposits of similar type. Bodies that are completely concealed may be included if there is specific geologic evidence of their presence." (Bureau of Mines and Geological Survey, 1948, p. 15.)

It follows that the whereabouts of reserves must still be established and their size and quality proved by the customary detailed exploration and evaluation undertaken by the industry. However, the information provided by this survey should assist in the selection for the best targets for such further work.

The following arbitrary physical criteria have been adopted:

a The deposit should average at least 1.0 m in thickness.

b The ratio of overburden to sand and gravel should be no more than 3:1.

c The proportion of fines (particles passing the No. 240 mesh BS sieve, about $\frac{1}{16}$ mm) should not exceed 40 per cent.

d The deposit should lie within 25 m of the surface, this being taken as the likely working depth under most circumstances. It follows from the second criterion that boreholes are drilled no deeper than 18 m if no sand and gravel has been proved.

A deposit of sand and gravel which broadly meets the above criteria is regarded as 'potentially workable' and is described and assessed as 'mineral'.

For the particular needs of assessing sand and gravel resources, a grain-size classification based on the geometric scale $\frac{1}{16}$ mm, $\frac{1}{4}$ mm, 1 mm, 4 mm, 16 mm etc has been adopted. The boundaries between fines (that is, the clay and silt fractions) and sand, and between sand and gravel material, are placed at $\frac{1}{16}$ mm and 4 mm, respectively 'see Appendix C).

The characteristics of the sand and gravel are assessed within resource blocks and generally, deposits thought to be of the same formation (and of approximately the same age) throughout the assessed area are given the same block letter, whether they contain potentially workable sand and gravel or not; for example block F embraces all of the First Terrace deposits (Table 4). Separate parts of a block which contain mineral are designated as sub-blocks and identified by a subscript figure, for example F_1.

Resource block O, the exception to this scheme, is an omnibus block which, in this assessment area, contains scattered usually minor occurrences of glacial and river terrace sand and gravel and major areas of Jurassic bedrock.

In the assessment of mineral no account is taken of factors such as roads, villages and land of high agricultural or landscape value, which might stand in the way of sand and gravel exploitation, although towns are excluded. The estimated total volume of mineral, therefore, bears no simple relationship to the amount that could be extracted in practice.

It must be emphasised that the quantitative assessment applies to the mineral in a sub-block as a whole. Valid conclusions cannot be drawn about parts of the mineral area except in the immediate vicinity of the actual sample points.

DESCRIPTION OF THE RESOURCE SHEET

GENERAL

Sheets TF 00 and TF 10 depict 200 km^2 of country lying between Stamford [030 070] in the north-west and Greater Peterborough [180 020] in the south-east (Figure 1). Some 61.0 km^2 of this area is mineral-bearing (Table 2); other sand and gravel deposits within Greater Peterborough (see the map) have not been assessed.

TOPOGRAPHY

The area is readily divided into two physiographical regions, the Upland (that is in the sense of Seale, 1975), and the Fenland. The latter lies in the northern and eastern parts and includes the major part of the mineral-bearing ground (Figure 2).

The Upland, a dissected plateau with mature woodlands set in arable, hedge-lined fields, is underlain mainly by Lower and Middle Jurassic clays and limestones. The ground is generally over 30 m (100 ft) above Ordnance Datum and rises gently westwards, reaching a maximum of over 91 m (300 ft) above Ordnance Datum near Easton on the Hill [013 043].

By contrast, the Fenland is usually below 10 m (30 ft) Ordnance Datum and is relatively flat; hedges are rare or absent and fields are commonly outlined by a rectilinear network of ditches which feed into major drainage channels. The agricultural value of the land is high; it is used either for livestock rearing or intensive arable cropping of potatoes, sugar beet, other root crops and cereals. Trees are usually restricted to hedgerows and the vicinity of farm buildings. The soils are developed either from river terrace deposits or from silts, clays and peats, particularly along the north-eastern margins of the area.

The drainage of these two regions is dominated by the eastward-flowing River Welland. From the confines of a narrow valley within the Upland, the river emerges onto the broad, gently inclined Fenland depositing alluvium over older, fan-like spreads of gravel.

In addition, several minor streams, notably the River Gwash, flow into the Welland below Stamford, and an unnamed misfit tributary flows south-westwards to join the River Nene (off the map).

GEOLOGY

Introduction and previous work

The area was originally geologically surveyed on the one-inch scale by J. W. Judd, W. H. Holloway and S. B. J. Skertchly and published on the Old Series Sheet 64 in 1872 (Solid) and 1877 (Drift). The area, which lies west of grid-line easting 12, was surveyed on the six-inch scale between 1939 and 1947 by G. A. Kellaway, J. H. Taylor, W. B. Evans and G. Bisson and published as part of the Stamford (157) One Inch New Series Geological sheet in 1957. A revised edition of this map on the scale of 1:50 000 was published in 1977.

Additional six-inch mapping by A. Horton, R. D. Lake and B. C. Coppack during 1968 was undertaken for a survey of the Peterborough New Town area and this was published on the scale of 1:25 000 together with a descriptive account of the geology (Horton and others, 1974). The remainder of the ground included in the present assessment was mapped on the six-inch scale by R. J. Wyatt in 1977.

The following account is based on information provided by members of the Institute's East Anglia and South-Eastern England Field Unit, notably R. J. Wyatt. The deposits are listed—see Table 1, and described in order of decreasing age.

Structure

The outcrops of the solid rocks shown on the map range from Upper Lias to Oxford Clay of Lower, Middle and Upper Jurassic age (Table 1). Non-IMAU boreholes in the area have proved strata older than the Upper Lias but consideration of them is unnecessary for the purposes of this assessment.

The geological structure is relatively simple. The regional dip is uniformly to the east-south-east and does not exceed 5 degrees. The only important disturbance is the west-north-west-trending Tinwell–Marholm Fault, the downthrow of which is estimated at 15 to 30 m to the north. Associated with this fault

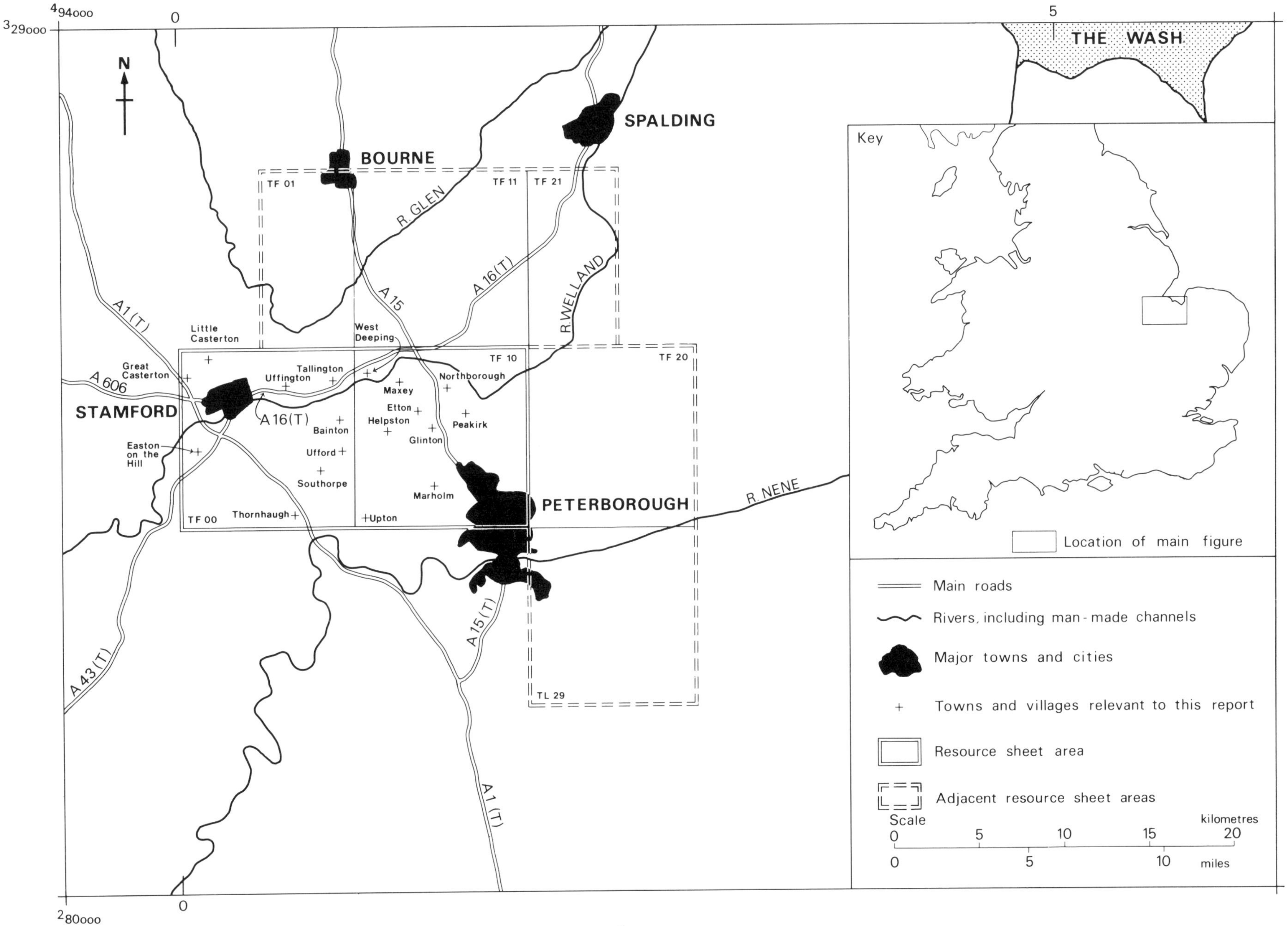

Figure 1 Locality map

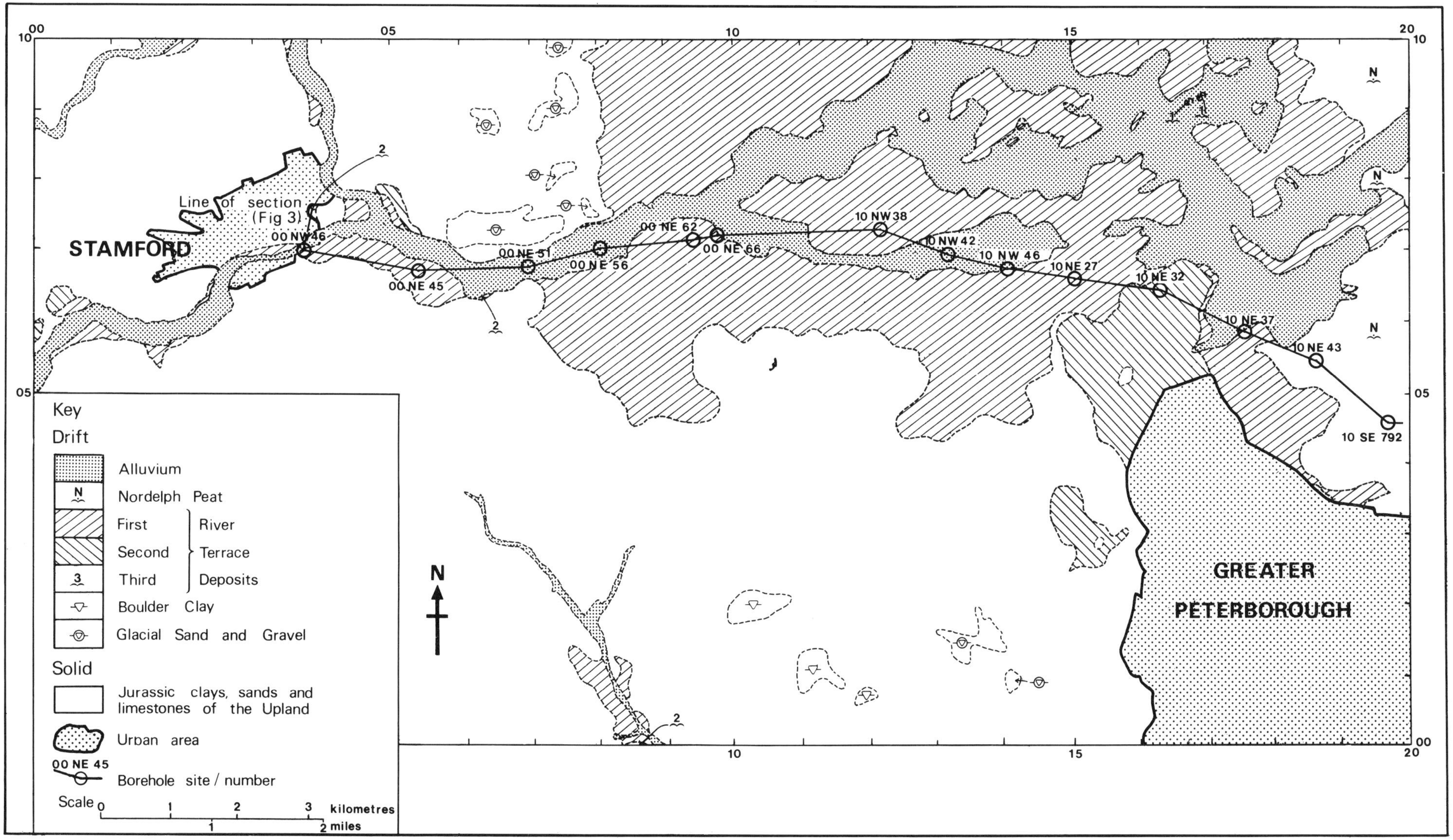

Figure 2 Drift deposits in the area

Table 1 Geological succession proved at the surface and in IMAU boreholes

			Thickness (recorded range in metres)
DRIFT			
Quaternary			
Recent and Pleistocene	*Flandrian Deposits*		
	Alluvium (freshwater)		0.4–4.2
	Nordelph Peat (Upper Peat of some authors)		0–0.7
	Barroway Drove Beds (estuarine and marine alluvium)		0–1.7
	River Terrace Deposits		
	First Terrace, including an upper clayey deposit—the 'Crowland Bed'	mapped as 'Fen Gravel(s)' in area of Stamford (157) sheet and also on Peterborough 1:25 000 sheet (Horton and others, 1974)	up to 7.1
	Second Terrace		up to 3.5
	Third Terrace		rarely exceeds 2
	Head		up to 4.5
	Boulder Clay		up to 12.2
	Glacial Sand and Gravel		up to 4.2
SOLID			
Jurassic			
Upper	Oxford Clay (Lower and Middle)		up to 33
	Kellaways Beds (includes Kellaways Sand and Kellaways Clay)		4.3–6.7
	Cornbrash (in part)		
Middle	Cornbrash (in part)		1.4–2.4
	Blisworth Clay		3.0–6.0
	Blisworth Limestone		2.8–4.6
	Upper Estuarine 'Series' (including Upper Estuarine Limestone)		5–11
	Lincolnshire Limestone (including the Upper and Lower members separated by the *crossi* Bed)		18.0–24.5
	Lower Estuarine 'Series'		1.5–6.0
	Northampton Sand		2.5–6.0
Lower	Upper Lias		45–52

between Ufford [094 042] and Marholm [148 021] are slight flexures, particularly on the north side, and small faulted blocks on the south side. Relative uplift of the strata on the south side of the fault has resulted in an eastward displacement of the Middle Jurassic outcrops. Subsequent erosion of the relatively less resistant Oxford Clay on the north side of the fault has produced an embayment of low ground now covered with Drift deposits.

Of the minor faults only one trending north-west to south-west through Tallington [095 085] reaches the Fenland and this has an estimated maximum downthrow to the north-east of 9 m. (Note: the sub-drift geological boundaries adjacent to this fault shown on the published 1:50 000 Stamford Sheet (157) have been re-interpreted in the light of information from IMAU boreholes and incorporated on the map accompanying this report.)

In addition to the tectonic structures, superficial disturbances characterise the valley slopes and floors (Hollingworth, Taylor and Kellaway, 1944; Horswill and Horton, 1976). Thus, cambering on the slopes and fractures parallel to the valley sides are common, whilst bulging has disrupted and forced up the clays of the Upper Lias and superincumbent strata in the valley floors. Landslipping of the Upper Lias clays has also occurred on the sides of the Welland valley near Easton on the Hill.

Stratigraphy

SOLID

The greater part of the solid outcrop is occupied by relatively durable shelly and oolitic limestones which are prominent in the gravel component of the local drift deposits (see Figure 4).

Upper Lias The outcrop of the Upper Lias is restricted to the valleys of the River Welland and other minor streams near Stamford, and to Burghley Park [047 055] on the upthrow (southern) side of the Tinwell–Marholm Fault.

The formation consists of grey clays and mudstones with thin limestone bands in the lowest 3 m. It is normally 45 to 52 m thick but may be substantially reduced in thickness beneath valley slopes because of cambering.

Northampton Sand The Northampton Sand overlies the Upper Lias and its outcrop is similar in shape and extent. The formation consists of ironstones, sandy sideritic limestones and sideritic mudstones. South-east of Burghley Park, the formation becomes very thin, although boreholes indicate that it persists at least as far as Greater Peterborough where it comprises less than 1.0 m of bioclastic limestone interbedded with rootlet-bearing sandy clays.

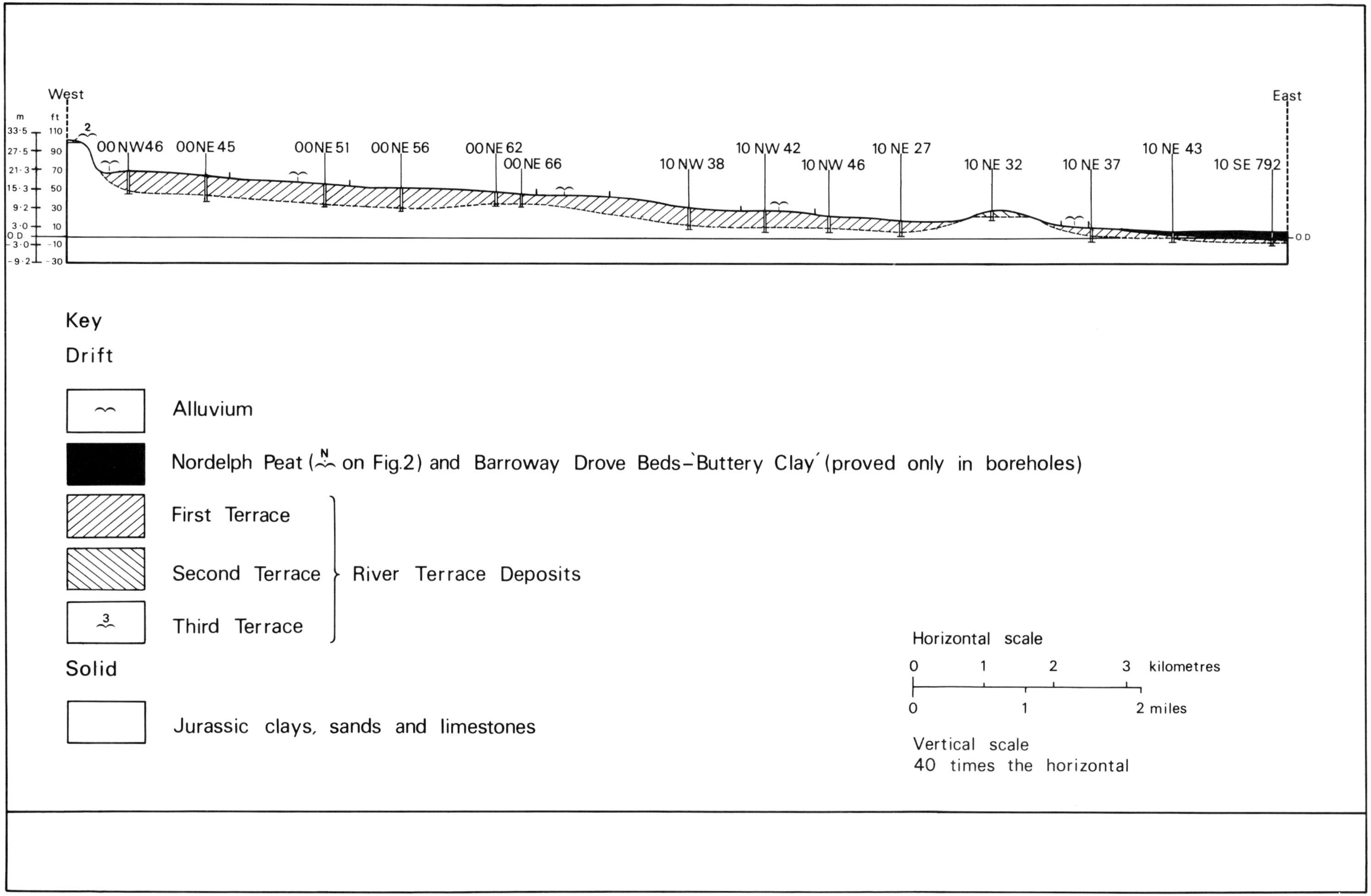

Figure 3 Section through the Fenland (for line of section see Figure 2)

Lower Estuarine 'Series' Fine sands, silts and silty clays, including rootlet beds and seatearths characterise the Lower Estuarine 'Series' whose outcrop is often obscured by an extensive downwash of Lincolnshire Limestone debris.

Lincolnshire Limestone The Lincolnshire Limestone has a wide outcrop both north and south of Stamford and forms extensive upland plateaux to the south and south-east of the town, with an elevation generally 76 m (250 ft) above Ordnance Datum, but not usually exceeding 95 m (312 ft) above Ordnance Datum. In these areas the limestone is 18 to 21 m thick with a maximum of 24.5 m proved in the Glinton borehole [1502 0526]. However, towards the south-east the limestone thins rapidly and disappears along a line running south-west to north-east through Peterborough.

Within this formation, upper and lower divisions are recognised: north of the Welland valley they are separated by a thin bed of limestone distinguished by an abundance of the brachiopod *Acanthothiris crossi* which forms a readily mappable unit overlying the Lower Lincolnshire Limestone. South of the Welland valley the *crossi* Bed and overlying Upper Lincolnshire Limestone are usually absent beneath the unconformity at the base of the Upper Estuarine 'Series' (Hains and Horton, 1969, pp. 83–84). However, the Upper Lincolnshire Limestone is locally preserved in channels cut into the Lower Lincolnshire Limestone (Taylor, 1947).

The Lower Lincolnshire Limestone consists of level-bedded oolites and oolitic limestones with fine-grained sandy limestones in the lower part, the latter often being decalcified into loose sands. At the base, near Stamford, the sandy limestones are fissile and constitute the Collyweston Slate. The Upper Lincolnshire Limestone comprises coarse, shelly, cross-bedded oolites and freestones.

Upper Estuarine 'Series' These beds consist of a lower freshwater sequence of white siliceous clays, silts and sands showing the characteristics of fireclay or ganister, and an upper rhythmic sequence comprising alternations of laminated or shelly clays and clays with rootlets and/or comminuted plant debris.

At the base of the formation is a thin impersistent bed of nodular ironstone known as the Upper Estuarine Ironstone.

Blisworth Limestone On the Stamford (157) One-Inch sheet, and on some six-inch quadrants of 1:25 000 sheet TF 00, the Blisworth Limestone and overlying Blisworth Clay are given their older names of Great Oolite Limestone and Great Oolite Clay, respectively. However, this report follows more recent Survey practice and they are referred to as the Blisworth Limestone and Blisworth Clay, respectively (for example the published 1:50 000 Stamford (157) sheet and the 1:25 000 Peterborough sheet).

The formation consists of blue-hearted, shelly, micritic limestones with minor beds of shelly marl and clay.

Blisworth Clay This formation consists of grey, green and dark purple-tinted grey and often brightly coloured clays with shelly bands and occasional layers of ironstone nodules.

Cornbrash The Cornbrash consists mainly of indurated, bioclastic limestone which is blue-hearted and massive when unweathered but reddish brown and rubbly at outcrop. It usually produces a flat, bench-like feature, for example in the Uffington [063 077] and Bainton–Etton [094 061 to 141 065] areas, and also between Upton [107 005] and Peterborough where it produces a more extensive plateau.

Kellaways Beds The Kellaways Beds comprise dark grey clays (Kellaways Clay) overlain by silts or fine-grained sands (Kellaways Sand) which are occasionally patchily cemented to form 'doggers'. These beds crop out south of the Tinwell–Marholm Fault on the higher ground north of Upton; more extensive occurrences on the northern, downthrow side of the fault between Uffington and Peterborough are present both at outcrop and beneath river terrace deposits.

Oxford Clay The Oxford Clay consists mainly of bluish grey and greenish grey mudstone which weathers to a pale brown plastic clay and produces a heavy clay soil. The clay is completely decalcified at outcrop to a maximum depth of about a metre. The mudstone consists mainly of the clay mineral illite with subsidiary kaolinite. Thin bands of sandy argillaceous limestone or layers of septaria are common. Weathering of the abundant pyrites and calcium carbonate gives rise to numerous selenite (calcium sulphate) crystals in the superficial brown oxidised layers of the clay.

Three major lithological divisions broadly coinciding with three faunal divisions may be recognised, namely the Lower, Middle and Upper Oxford Clay. In the assessed area the Lower and Middle Oxford Clay occur either at outcrop or beneath River Terrace Deposits between Bainton and Peterborough. The Upper Oxford Clay is absent.

The Oxford Clay contains abundant ammonites together with bivalves (notably *Gryphaea*), brachiopods, gastropods, belemnites and annelids and is renowned for well preserved specimens of marine reptiles including plesiosaurs, ichthyosaurs and crocodiles (Horton and others, 1974, p. 50); fragments of invertebrate fossils (particularly *Gryphaea* and belemnites) are commonly found in the Glacial Sand and Gravel and River Terrace Deposits.

DRIFT

Pleistocene The most widespread Pleistocene deposits of this area are the river terrace sands and gravels. Relatively small, isolated patches of glacial material also occur.

Glacial Sand and Gravel Patches of Glacial Sand and Gravel occur at and north of Uffington and also near Marholm. The deposits at Uffington are thought to be glacial because of their close association with Boulder Clay north of the village (on 1:25 000 sheet TF 01), and those south-west of Marholm because of their high topographical position. The deposits contain many pebbles of subrounded oolitic limestone, and angular to subangular flint with quartzite and ironstone (see Composition of the Sand and Gravel). Of the outcrops investigated, only those at Uffington and Marholm are assessed as mineral deposits.

Boulder Clay This is the 'Chalky Boulder Clay' of Horton and others, (1974, p. 51) and the 'Chalky/Jurassic till' of Gallois, (1979, p. 32.) Small outcrops of Boulder Clay occur west of Thornhaugh [067 005] and at Upton. When unweathered this comprises stiff grey clay with abundant rounded to angular chalk and flint clasts with subrounded to subangular 'Bunter' quartzite pebbles, far-travelled erratics (Sabine, 1949) and locally-derived rock debris; upon exposure the clay weathers greyish brown to dark brown. The maximum thickness (12.2 m) was recorded in borehole 00 SW 104 [0275 0000].

Head Small, localised deposits of Head occur on the Upland where they generally occupy valley floors, for example, south of Casewick Hall at [078 085], at Southorpe [082 031] and at Little Casterton [021 098]. At the last two localities the data from boreholes 00 SE 2 and 00 NW 55 suggest that Head overlies River Terrace Deposits.

Because Head deposits have accumulated through soil creep (or solifluction) their composition varies widely depending upon the local parent rock, and they include loams, stony-clays and unsorted rock debris. Generally, Head deposits are about 1 m thick but a maximum of 4.5 m was recorded in borehole 00 NW 55.

River Terrace Deposits *Third Terrace* deposits of the Nene underlie about one-third of the Greater Peterborough conurbation included in the map. They comprise sandy clay and gravel, the latter often very clayey. The deposits, which rarely exceed 2 m, are extensively cryoturbated and admixed with the top of the underlying Oxford Clay or Kellaways Beds. The surface of the terrace falls gently towards the north-east from 14 to 13 m above Ordnance Datum. No assessment of these deposits has been presented.

Sand and gravel deposits, other than those of the Third Terrace and others of presumed glacial origin, were formerly grouped together as *'Fen Gravel(s)'* (Skertchly, 1877, p. 184; the New Series One-Inch Geological sheet 157; the 1:50 000 Stamford (157) sheet; and Horton and others, 1974, pp. 57–58).

The present survey, together with the Author's and R. J. Wyatt's studies east of Peterborough have shown that locally, the former 'Fen Gravel(s)' can be sub-divided on the basis of height differences into first and second river terraces.

Small, isolated outcrops of *Second Terrace* occur in the vicinity of Stamford [at 001 059, 004 064, 052 078 and at 0648 0638] comprising clay with some sand and a trace of gravel.

Two more extensive areas of Second Terrace in the Glinton–Peakirk [168 067] area and also north of Marholm are assessed as mineral-bearing. They generally comprise rather clayey gravels containing flints and quartzite pebbles, occasional cobbles of 'Bunter Sandstone', and fragments of locally derived oolitic and shelly limestone with ironstone. The maximum thickness recorded was 3.5 m, with a mean of 1.3 m.

First Terrace deposits were derived principally from the outwash of the River Welland and its tributary, the Gwash, and constitute the major mineral-bearing stratum in the assessed area.

Outside the Fenland depositional basin, there are patches of First Terrace deposits flanking and underlying the Alluvium of an unnamed stream, near Southorpe, which drains southwards into the River Nene.

Much of the First Terrace sand and gravel of the Welland crops out along a broad north–south belt adjacent to the Upland area. It persists eastwards beneath the Recent deposits of the Fenland. The drying out and subsequent wastage of the former peat cover in post-Roman times has exposed additional outcrops of First Terrace (see Nordelph Peat).

Thickness variations within the First Terrace deposits suggest that they were formed as coalescing fans of gravel deposited by rivers debouching from steep-sided valleys in the Jurassic Upland onto the broad, low-lying Fenland.

Over most of the First Terrace outcrop within the Fenland there is a distinct differentiation between ubiquitous basal gravels and sandy gravels and a discontinuous relatively thin overlying spread of slightly gravelly loamy sands, silts and silty clays. The latter, which reach a recorded thickness of 1.7 m in boreholes 10 NE 26 and 10 SW 10, are interpreted either as representing the final stages of First Terrace aggradation—an older alluvium (personal communication, A. Horton and R. J. Wyatt) or as a residuum resulting from weathering of the terrace (personal communication, A. J. Dixon).

Beneath the Flandrian deposits of the Fenland the gravels are overlain by a thin heterogenous deposit of silty, sandy clays and clayey sands containing scattered pebbles. This deposit, here called the 'Crowland Bed' (and for simplicity included with First Terrace), appears to correspond stratigraphically with the 'older alluvium' noted above, but its mode of origin remains uncertain.

Organic clay lenses in the basal gravels at the Maxey pits [125 078] have yielded molluscs such as *Valvata piscinalis* (Müller), and *Pisidium henslowanum* (Sheppard) (identified by D. K. Graham of the Palaeontology Unit), which indicate a freshwater environment.

The First Terrace deposits have a mean thickness of 4 m with a maximum thickness of 7.1 m recorded in borehole 00 NW 55 (for further compositional data, see Composition of Sand and Gravel).

Recent These deposits include the peats, silts and clays of the Fenland basin (Godwin and Clifford, 1938–1940; Willis, 1961).

Barroway Drove Beds (Gallois, 1979) These beds are not exposed in the assessed area but were proved in boreholes 10 NE 43 and 10 SE 292 (see Figure 3). They generally consist of soft, wet, bluish grey clays interbedded with occasional silt laminae, often containing carbonaceous root traces and scattered shell fragments; these beds are the 'buttery clay' of earlier authors (for example Skertchly, 1877, p. 173), so-called because of their thixotropic properties.

These clays probably represent a salt-marsh deposit. Within the assessed area a maximum thickness of 1.7 m was recorded but much greater thicknesses have been observed farther east towards the centre of the Fenland basin.

Nordelph Peat (Gallois, 1979) The Nordelph Peat (or Upper Peat of some authors, for example Skertchly, 1877, p. 128) occurs only along the eastern margin of the assessed area and has a slightly more extensive distribution than the underlying Barroway Drove Beds. Formerly, the peat extended farther westwards but it has gradually been lost from these areas by shrinkage and erosion since the large-scale drainage of the Fenland became established in the mid-seventeenth century under Cornelius Vermuyden (Darby, 1940; Fillenham, 1963; Fowler, 1933; Godwin, 1978; Richardson and Smith, 1977; and Robinson, 1968).

The peat comprises mainly reed and sedge remains with alder brushwood in places. Tree trunks (locally known as 'bog oaks'—Seale, 1975, p. 7) are commonly ploughed up from the present-day peat/soil layers. The peats give rise to the dark humic soils of the 'Black Fens' (Astbury, 1958). Original maximum thicknesses are not known but a maximum of 0.7 m was recorded in borehole 10 NE 46 during the present survey.

Alluvium Alluvium floors the valley of the River Welland and extends out into the Fenland, forming a widespread but irregular cover to the underlying First Terrace deposits. The alluvial deposits are, in part, contemporaneous with the Barroway Drove Beds and Nordelph Peat but, in the extreme east, the most recent alluvial clays overlie the latter and are thus the youngest deposits in the Fenland sequence.

Generally, the Alluvium comprises mottled greyish brown clay with plant debris and sometimes pebbles; occasionally, and usually in deep channels, there are also beds of organic silt, peat and lenses of sand and gravel. These deposits are usually between 0.4 and 3.5 m thick in the Fenland but greater thicknesses are recorded locally in tributaries within the Upland; for example a recorded maximum of 4.2 m occurs in borehole 00 NW 53.

COMPOSITION OF THE SAND AND GRAVEL

The potentially workable sand and gravel deposits in the assessed area are Glacial Sand and Gravel and River Terrace Deposits: the latter constitute the major mineral resource.

Because both deposits are broadly similar in composition, the following account, unless otherwise stated, makes no distinction between them (see Figures 4 and 6 for summary information; for local variations see Tables 5 to 17 and Figures 5 and 7).

Limestone, the most common component of the gravel fraction, exceeds 80 per cent of the deposit in some localities (for example borehole 10 SE 786) and rarely falls below 30 per cent (for example borehole 00 NW 58). The range of variation in the mean values for the sub-blocks is 45 to 58 per cent. Flint occurs next in frequency (12 to 28 per cent) although occasionally it exceeds limestone (for example in boreholes 10 NW 31 and 10 NW 46 with a limestone:flint ratio of 21:71 and 16:69 per cent, respectively). Ironstone is ubiquitous and although it is generally represented in smaller quantities than flint, it ranks as a major constituent (8 to 28 per cent). Exceptionally it may represent up to 70 per cent of the deposit (for example borehole 00 NW 58).

Subsidiary components include quartzites, sandstones, derived fossil fragments and igneous

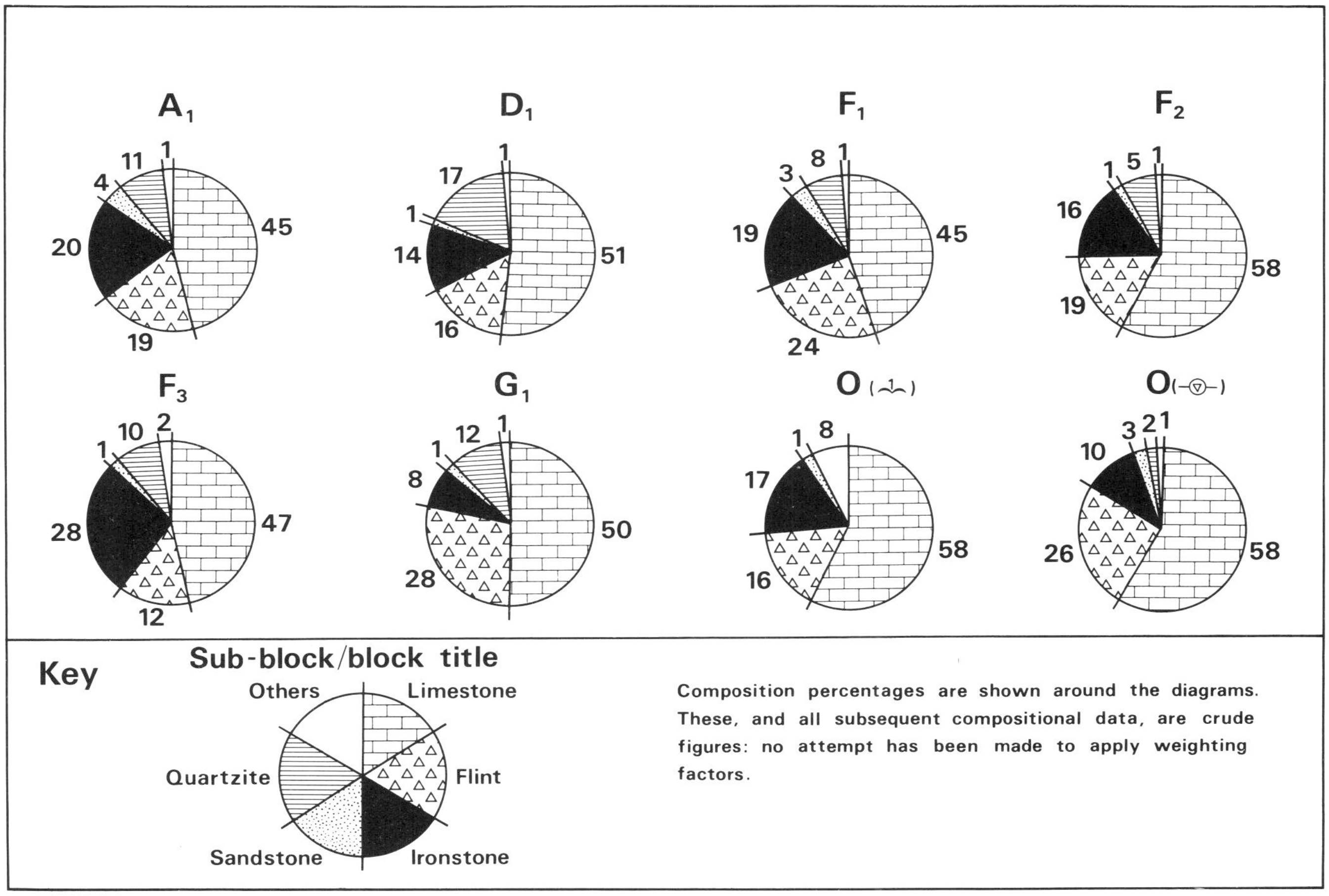

Figure 4 Mean composition by weight of gravel (+4–64 mm) in sub-blocks A_1 to G_1 and block O

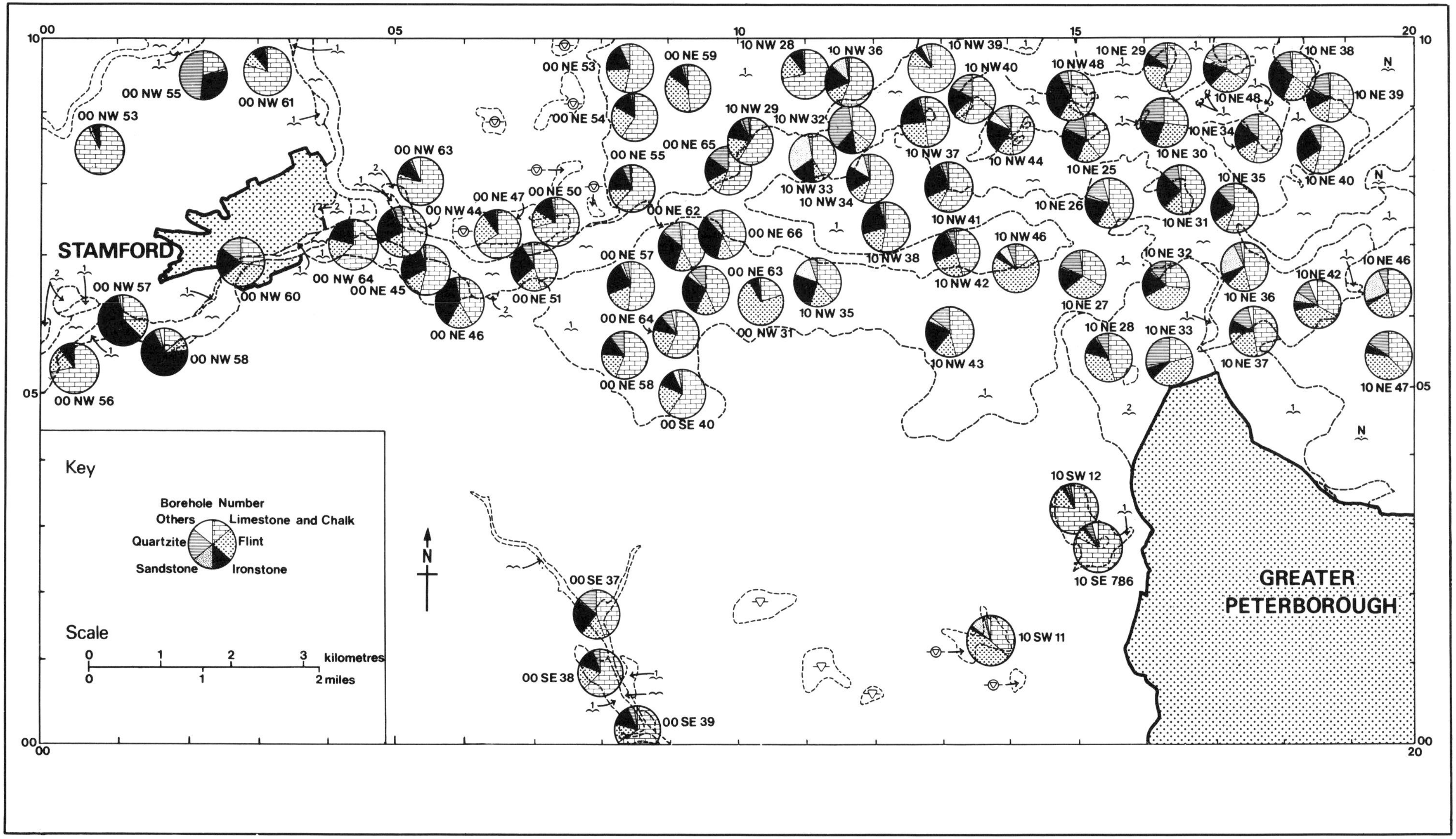

Figure 5 Mean composition by weight of gravel (+4–64 mm) in *some* IMAU boreholes

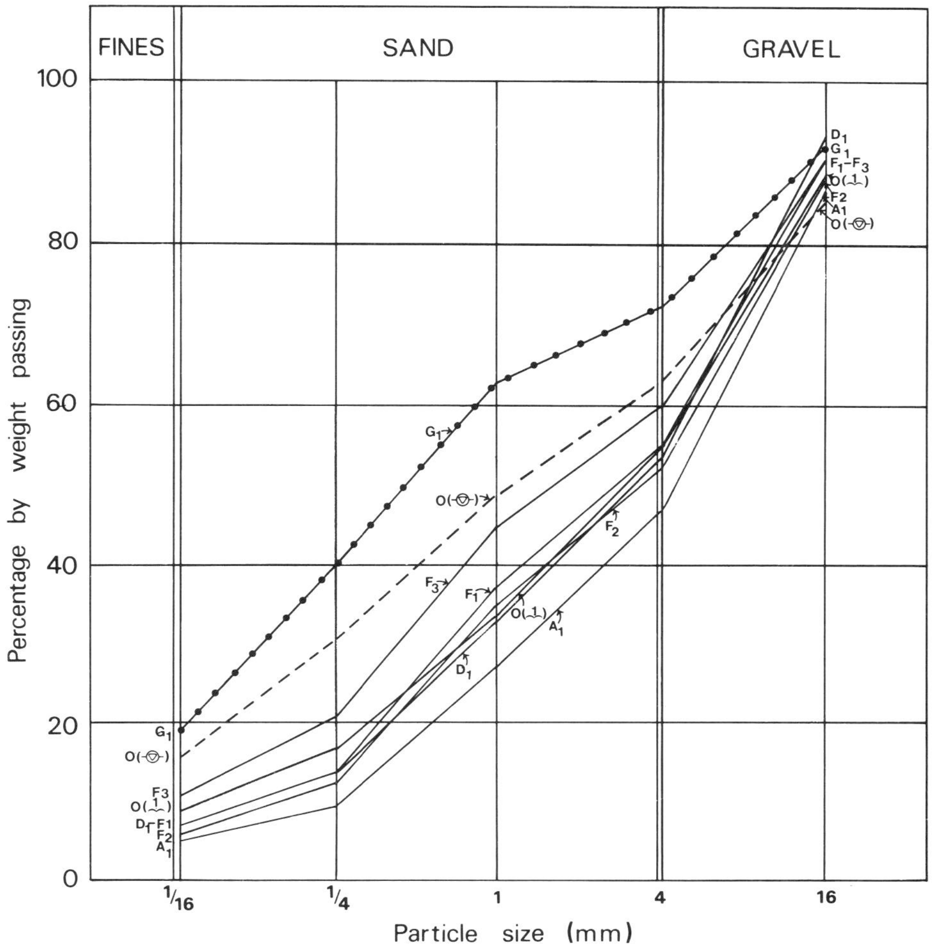

Key: for diagram above

•—•—• Second Terrace (Nene (G_1))

- - - - Glacial Sand and Gravel (O(-⊗-))

—— First Terrace (exposed or beneath cover) (Welland (A_1-D_1-F_1-F_2-F_3) and Nene tributary (O(⏊)))

Sub-block/ block	Percentage by weight passing				
	1/16mm	1/4mm	1mm	4mm	16mm
A_1	5	9	27	47	87
F_1	7	14	37	55	91
F_2	6	13	35	52	88
F_3	11	21	45	60	91
G_1	19	40	63	73	93
D_1*	7	14	33	54	94
O(⏊)	9	17	34	55	89
O(-⊗-)	16	31	49	63	86

Note: * refers to borehole 10 NE 39 only

Figure 6 Mean particle size distribution for the assessed thickness of mineral in sub-blocks A_1 to G_1 and block O

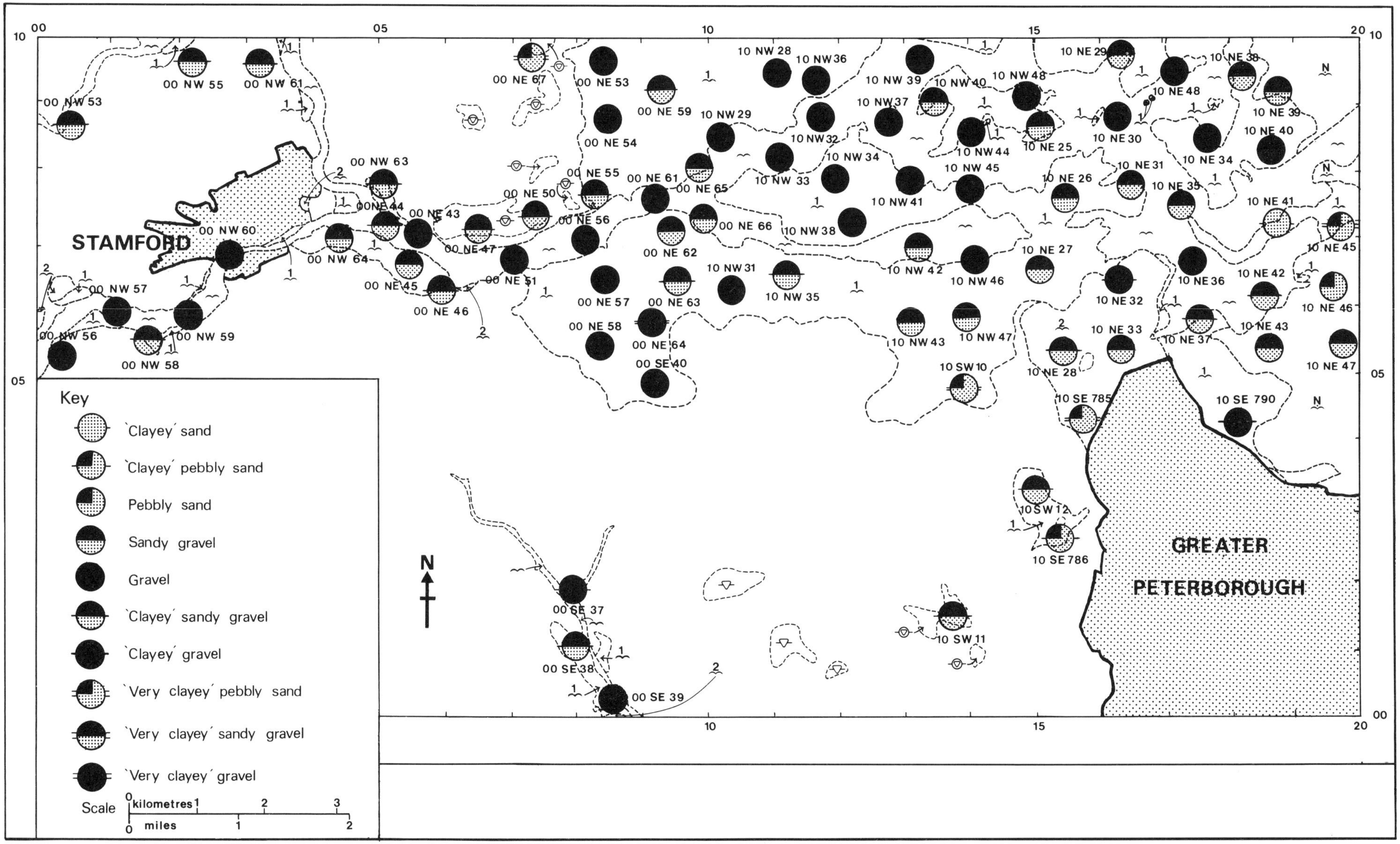

Figure 7 Mean grading characteristics of the sand and gravel

erratics; of these quartzite is predominant. Of the minor constituents not distinguished separately, chalk is included with the limestone as the frequency of chalk pebbles is extremely low.

The limestone pebbles are either shelly or oolitic and are derived from the Jurassic rocks outcropping extensively within the Upland and to the west and south of the assessed area. They are generally subangular to subrounded and often tabular.

Some limestone pebbles are veneered with reprecipitated iron oxide. Iron staining is also a common characteristic of flint gravel.

Flint (thought to have been derived from a former widespread Boulder Clay cover) is usually angular to subangular, generally white or brown, although black pebbles also occur. No quantitative assessment was made of the possibly deleterious amounts of white flint (Roeder, 1977; Figg, 1977).

Ironstone is derived principally from the sideritic facies of the Northampton Sand (Taylor, 1949, pp. 1–2), which crops out on the higher ground above the Welland valley and west of the assessed area. The pebbles are generally of fine gravel grade (+4–16 mm), irregularly shaped, moderately rounded and vary in colour from black to dark red dependent on the iron:quartz ratio of the ironstone.

The source of the rounded to subrounded quartzites and sandstones ranging from pebble to large cobble (maximum recorded dimension 0.3 m) is less certain although they probably originate from the Sherwood Sandstone Group (the 'Bunter Pebble Beds' and the 'Bunter Sandstones' of the literature) of the Midlands. Their initial transportation into the district was probably within the Boulder Clay during an ice advance, before a later fluvial reworking and subsequent deposition.

Material in trace amounts includes derived fossil fragments, commonly from the Oxford Clay, for example bivalves (*Gryphaea*), belemnites and occasional ammonites, often impregnated with iron pyrites. Occasional dark, fine-grained igneous or metamorphic erratics are found (Sabine, 1949). The scarcity of chalk (previously noted) is highlighted by its comparative abundance in gravels both north (around Witham, Lincolnshire) and south (in Cambridgeshire) of the assessed area (personal communication, G. Power and A. J. Dixon). In these latter areas the interfluves are formed partly by chalk escarpments; however, within the River Welland catchment area no chalk bedrock is exposed and its only source here is the Boulder Clay. However, any chalk derived from this source would probably rapidly disintegrate and be carried in solution by the time it reached the Fenland floodplain.

Although there is little compositional variation throughout the assessed area, the mean grading readily distinguishes a comparatively higher fines and sand content in the Glacial Sand and Gravel (in block O) and the Second Terrace deposits of the River Nene (sub-block G_1) compared with those of the First Terrace deposits of the River Welland (sub-blocks A_1 to F_3) and the unnamed tributary of the River Nene (in block O—see Figure 6).

The former deposits are 'clayey' sandy gravels whereas the mean gradings of the First Terrace deposits exhibit little variation, most being within the gravel or sandy gravel categories. Locally, there are wide variations between boreholes in the same deposit and sub-block (for example in sub-block F_1 borehole 10 SW 10 has a fines:sand:gravel ratio of 37:55:8 compared with borehole 10 NW 31 with a ratio of 6:46:48).

THE MAP

The sand and gravel resource map is folded into the pocket at the end of this report. The base map is a specially prepared composite Ordnance Survey 1:25 000 Outline Edition in grey, on which the topography is shown by contours in green, the geological data in black and the mineral resource information in shades of red.

Geological data The geological boundary lines and symbols, derived from previous surveys, represent the best interpretation of the information available at the time of the survey. However, it is inevitable, particularly with drift deposits which change rapidly vertically and laterally, that local discrepancies may occur.

Borehole data, which include the stratigraphical relations and mean particle size distribution of the sand and gravel samples collected during the assessment are also shown on the map.

Mineral resource information For assessment purposes the map is divided into resource blocks within which areas of mineral are identified as sub-blocks (for definitions of 'mineral', resource blocks and sub-blocks see pp. 1 and 14).

Where 'mineral' is shown it is subdivided into one of two categories: 'exposed', where the thickness of overburden, commonly only soil and subsoil, averages less than 1.0 m thick, and 'continuous or almost continuous spreads beneath overburden'. However, within both these categories, there may be small patches where sand and gravel is absent or not potentially workable, for example around boreholes 00 NE 60 and 00 NW 59. Uncoloured parts of the map show bedrock outcrops, areas of non mineral-bearing superficial deposits and sand and gravel which is not potentially workable. Areas of unassessed sand and gravel in built-up areas are stippled red.

The distribution of these categories and the resource assessment are based on the mapped geological boundaries. No quantitative account is made of mineral transgressing these boundaries as feather-edge deposits, for example, along the boundaries between sub-block A_1 and blocks D and F (north-east of Greater Peterborough).

Worked areas and made ground The approximate extent of known worked-out ground (including that for sand and gravel) are shown to 1977; areas which have been backfilled are depicted as 'made ground' (for further details see Table 3).

RESULTS

The statistical results are summarised in Table 2. Additional summary compositional data are shown in Figures 4 and 6.

Accuracy of the results For the five resource sub-blocks where the volume assessment is calculated at the *indicated* level (that is in sub-blocks A_1, $F_{1/2/3}$ and G_1 extending over 57.1 km^2) the confidence limits at the symmetrical 95 per cent probability level vary

Table 2 Summary of results: the sand and gravel resources of the assessed area

Resource sub-block	No. of sample points	Area (km^2) Sub-block	Area (km^2) Mineral	Mean thickness (m) Over-burden	Mean thickness (m) Mineral	Volume of mineral (million m^3)	Limits at the 95 per cent confidence level ±%	Limits ±Volume million m^3	Mean grading percentage (based on IMAU data) Fines $-\frac{1}{16}$ mm	Sand $+\frac{1}{16}$ 4 mm	Gravel +4 mm
a ASSESSMENT OF SUB-BLOCKS A_1 TO G_1 AT THE 'INDICATED' LEVEL											
A_1	37	22.1	21.9	1.6	2.9	63.5	20	12.7	5	42	53
F_1	30	19.4	18.1	0.8	3.2	57.9	24	13.9	7	48	45
F_2	16	7.0	6.0	0.4	3.5	21.0	26	5.5	6	46	48
F_3	20	7.4	6.9	1.0	3.3	22.8	20	4.6	11	50	39
G_1	14	4.4	4.2	0.7	1.3	5.5	39	2.1	20	53	27
A_1–G_1	117	60.3	57.1	1.1	3.0	170.7	11	18.8	7	46	47
b ASSESSMENT OF SUB-BLOCKS D_1 AND BLOCK O AT THE 'INFERRED' LEVEL											
i First Terrace (Welland) beneath Nordelph Peat											
D_1	8	6.9	1.7	1.2	2.8	4.8	Speculative		7	47	46
ii First Terrace (Nene tributary) partially obscured by Alluvium											
O	4	112.9	1.1	1.3	6.0	6.6	Speculative		9	45	46
iii Glacial Sand and Gravel											
O	4	112.9	1.1	0.7	2.2	2.4	Speculative		16	47	37

between 20 and 39 per cent (that is, it is probable that 19 times out of 20 the true volume lies within the given limits). However, the true values are more likely to be nearer the figures estimated than the limits. Moreover, it is probable that in each sub-block roughly the same percentage limits would apply for the estimate of volume of a very much smaller parcel of ground (say 100 hectares) containing similar sand and gravel deposits if results from the same number of sample points (as provided by, say 10 boreholes) were used in the calculation. Thus, if closer limits are needed for the quotation of reserves of part of a sub-block, it can be expected that data from more than 10 sample points will be required, even if the area is quite small. This point can be illustrated by considering the whole of the potentially workable sand and gravel associated with the sub-blocks in the assessment area. The volume at the *indicated* level can be estimated to limits of ±11 per cent by a calculation based on data from 117 sample points.

For sub-block D_1, the Glacial Sand and Gravel and the River Terrace Deposits within block O an assessment is offered at the *inferred* level: confidence limits are not quoted (see Table 2b).

It must again be emphasised that the quoted volume of sand and gravel has no simple relationship with the amount that could be extracted in practice, as no allowance has been made in the calculation for any restraints (such as existing buildings and roads) on the use of the land for mineral working.

Worked-out ground: details The areas and estimated volumes of worked-out sand and gravel deposits are shown in Table 3. Some worked-out pits have been restored to ground level by infilling with refuse; others have been landscaped and allowed to fill with water for recreation and conservation purposes. It should be noted that on the map no distinction is made in the ornament for areas worked for sand and gravel and those worked for other bulk minerals, such as limestone.

Table 3 Areas and estimated volumes of worked-out sand and gravel deposits in the assessed area (shown to 1977)

Resource sub-block/block	Approximate area (km^2)	Estimated volume (million m^3)
A_1	0.2	0.6
D_1 ('inferred' assessment only)	—	—
F_1	1.3	4.2
F_2	1.0	3.5
F_3	0.5	1.7
G_1	—	—
O (First Terrace, 'inferred' assessment only)	—	—
O (Glacial Sand and Gravel, 'inferred' assessment only	0.1	0.2

NOTES ON RESOURCE BLOCKS

The block letters (A–I) follow the stratigraphical order of superficial deposits mapped in the Bourne [095 202], Stamford and Peterborough areas of which this report forms a part; for example block A includes the most recent deposits (Alluvium) and block I the oldest deposits (Third Terrace). Block O is the exception to this scheme—it contains scattered occurrences of glacial and terrace deposits (for further details regarding blocks, sub-blocks and subscript figures see pp. 2–3, Table 4 and Figure 8).

Letters B, C, H and I are not used in this report because the deposits to which they refer (see Table 4) do not crop out in the present map area. They will be used in later reports and are incorporated here in order to preserve the continuity between this resource scheme and subsequent publications concerned with the areas referred to above.

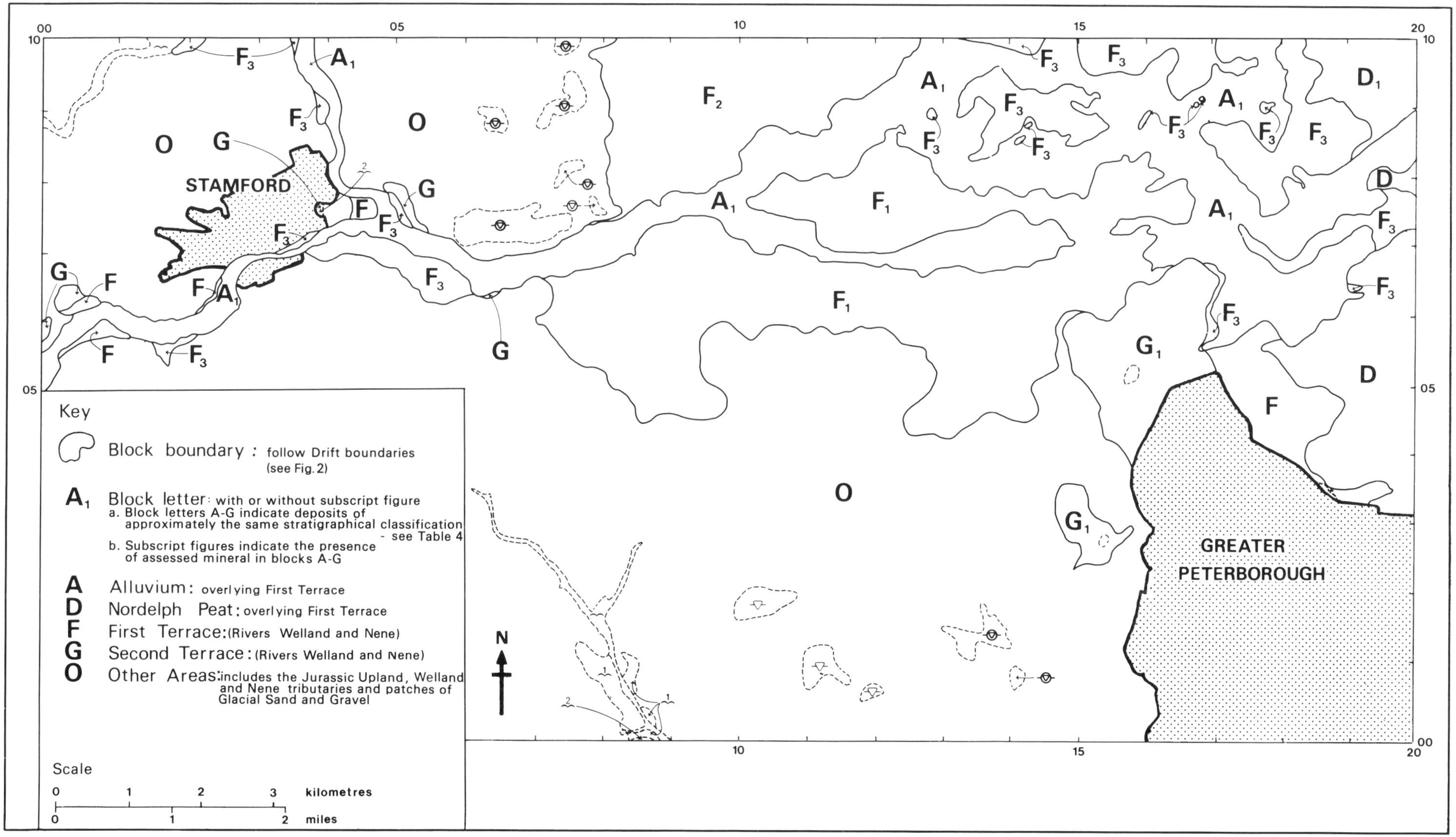

Figure 8 Location of resource blocks and sub-blocks

Table 4 The relationship of the block letters to the classification of the Drift deposits

Block letter	Drift deposits covered by each block
A	Alluvium—overlying First Terrace
B*	Shell Marl (Whittlesey area)
C*	Terrington Beds (Bourne area)
D	Nordelph Peat—overlying First Terrace
E*	Barroway Drove Beds—proved only in boreholes, e.g. 10 NE 43
F	First Terrace—Rivers Welland and Nene
G	Second Terrace—Rivers Welland and Nene
H*	March Gravels—(Whittlesey area)
I*	Third Terrace (Whittlesey area)
O	Other areas—including deposits of the Jurassic Upland, River Nene and Welland tributaries and patches of Glacial Sand and Gravel

*Note** Not used in this report.

The boundary of Greater Peterborough is taken from the Peterborough Development Corporation structure map published in 1970 (Anon., 1970; also Hancock and Hawkes, 1967).

Block A

This block encompasses the floodplain deposits of the River Welland including the lower reaches of a tributary, the River Gwash. The deposits, comprising Alluvial silts and clays overlying First Terrace sand and gravel, represent continuous or almost continuous mineral beneath overburden. The bedrock ranges from Upper Lias in the west to Oxford Clay in the east.

The block is 22.1 km^2 in area of which 21.9 km^2 is mineral bearing (sub-block A_1); sand and gravel has been extracted from the remainder which is shown as worked ground. The present-day workings are restricted to a small pit [050 071] which is being worked for roadstone and a larger pit [105 071] near Lolham Bridge.

Sub-block A_1 The assessment is based on data from 27 IMAU boreholes (of which four did not prove mineral), three Hydrogeology Unit borehole records and seven others. In boreholes 00 NW 59, 00 NW 60 and 10 NE 41, the sand and gravel did not meet the criteria a and/or b of the definition of mineral (see p. 1) but nevertheless the data have been used in the calculation of the resources.

The recorded thickness of overburden ranged from 0.4 to 3.5 m (in boreholes 00 NE 43 and 00 NW 59, respectively) with a mean thickness of 1.6 m. The greatest thicknesses mostly occur in the upper Welland valley where the overburden probably includes solifluction material. Mineral proven ranged from 0.4 to 6.1 m thick (in boreholes 00 NW 59 and 10 NW 41, respectively) with a mean thickness of 2.9 m.

Nineteen of the IMAU boreholes proved mineral graded as gravel, four proved sandy gravel, two proved 'clayey' sandy gravel and one 'clayey' sand. (Table 5 and Figure 7.) The mean grading data demonstrate that there is little variation between boreholes in the sub-block and that the deposits do not perceptibly change along the valley of the Welland, as might be expected.

The gravel (+4–64 mm) (Table 6) consists mostly of oolitic and shelly limestone (with a mean for the sub-block of 45 per cent) derived locally from the Blisworth and Lincolnshire Limestones, with ironstone (20 per cent) mostly from the Upper Estuarine Ironstone and Northampton Sand. Flint (19 per cent) is common and, together with quartzitic 'Bunter' pebbles, was probably derived from formerly more extensive Boulder Clay. Generally, within the sub-block the relative proportions of the constituents do not show a marked deviation from the mean. However, the upper limit of the ironstone (60 per cent) is attributed to one borehole (00 NW 57) which is located downstream from outcrops of Northampton Sand. Similarly, sandstone (4 per cent) and quartzite (11 per cent) also show relatively high upper percentages (33 per cent and 40 per cent, respectively), as a result of atypical occurrences in single boreholes.

The mean grading for the sub-block is gravel 53 per cent, sand 42 per cent and fines 5 per cent—with an overall classification as gravel. The volume of mineral is estimated to be 63.5 million $m^3 \pm 20$ per cent at the 95 per cent confidence level.

Block D (including D_1)

Three areas in the extreme east of the map, enclosing outcrops of Nordelph Peat, constitute block D which is 6.9 km^2 in area. Of this, only an isolated patch (1.7 km^2) in the north, around Stowgate Farm South [198 099] contains mineral and this lies beneath overburden. However, sand and gravel less than 1.0 m thick was proved in boreholes drilled in the largest part of block D, north-east of Peterborough. No borehole information is available for the smallest section of block D [195 080] but field observations suggest that mineral is absent. No records of gravel extraction within this block are known.

The assessment is based on six IMAU boreholes and two others. Of the IMAU boreholes only one (10 NE 39) proved mineral whereas boreholes 10 NE 43, 10 NE 46 and 10 NE 47 proved sand and gravel 0.6, 0.8 and 0.6 m thick, respectively. The remaining two boreholes (10 SE 791 and 10 SE 792) were barren.

Sub-block D_1 Because the area of mineral is less than 2.0 km^2 and there is information from only two boreholes (one of which was an IMAU borehole) an *inferred* assessment is offered (see p. 1 and Appendix B, paragraph 12).

The mean recorded thickness of overburden is 1.2 m. The proved mean thickness of the mineral is 2.8 m ranging from 2.5 m in IMAU borehole 10 NE 39 to 3.1 m in borehole 10 NE 8_A.

Borehole 10 NE 39 proved sandy gravel consisting mainly of fine with some coarse gravel (46 per cent), mainly medium to coarse sand (47 per cent) and fines (7 per cent). No other grading information is available. The gravel (+4–64 mm) comprises mainly oolitic and shelly limestone (51 per cent), with quartzite (17 per cent), flint (16 per cent), ironstone (14 per cent) and traces of sandstone and chalk—see borehole log for further grading and compositional details.

The mean grading for this sub-block is gravel 46 per cent, sand 47 per cent and fines 7 per cent—with an overall classification as sandy gravel. The *inferred* volume of mineral is estimated to be 4.8 million m^3.

Block F (including F_1, F_2 and F_3)

Block F includes all the First Terrace deposits of the

Table 5 Sub-block A_1: data from IMAU boreholes

Borehole No.	Recorded thickness		Mean grading percentage							Grading classification
	Overburden m	Mineral m	Fines $-\frac{1}{16}$ mm	Fine sand $+\frac{1}{16}-\frac{1}{4}$ mm	Medium sand $+\frac{1}{4}-1$ mm	Coarse sand $+1-4$ mm	Fine gravel $+4-16$ mm	Coarse gravel $+16$ mm		
00 NW 56	3.3	3.6	3	2	16	21	38	20		Gravel
00 NW 57	3.0	5.4	4	2	19	22	40	13		Gravel
00 NW 59	3.5	(0.4)§	6	4	13	13	31	33		Gravel
00 NW 60	2.8	(0.8)§	2	1	12	14	40	31		Gravel
00 NE 43	0.4	1.0	8	5	16	17	35	19		Gravel
00 NE 44	2.6	1.9	12	8	21	16	32	11		'Clayey' sandy gravel
00 NE 51	2.2	4.1	4	2	16	21	34	23		Gravel
00 NE 56	2.0	4.5	4	2	14	20	40	20		Gravel
00 NE 61	2.9	1.1	2	1	11	25	40	21		Gravel
10 NW 29	2.4	2.6	3	3	13	29	43	9		Gravel
10 NW 32	1.6	4.2*	4	4	17	23	44	8		Gravel
10 NW 33	2.3	1.2	4	3	16	19	39	19		Gravel
10 NW 37	1.1	6.0	4	4	23	18	38	13		Gravel
10 NW 39	0.9	3.1	3	2	15	21	47	12		Gravel
10 NW 40	1.0	2.2	3	7	24	20	40	6		Sandy gravel
10 NW 41	0.9	6.1	4	2	17	23	44	10		Gravel
10 NW 42	0.6	3.6†	6	5	25	18	38	8		Sandy gravel
10 NW 45	1.0	5.4	2	2	13	16	48	19		Gravel
10 NE 29	0.7	3.1	5	7	23	23	38	4		Sandy gravel
10 NE 30	1.8	2.7	5	5	19	23	39	9		Gravel
10 NE 34	1.0	2.8	5	6	23	16	39	11		Gravel
10 NE 35	1.2	1.3	8	8	26	23	29	6		Sandy gravel
10 NE 36	1.4	2.5‡	5	3	15	22	32	12		Gravel
10 NE 41	1.4	(0.8)§	18	32	42	6	2	0		'Clayey' sand
10 NE 42	1.2	1.7	11	7	22	16	38	6		'Clayey' sandy gravel
10 NE 44	Absent	Absent	—	—	—	—	—	—		—
10 NE 48	1.5	2.2	5	5	21	20	41	8		Gravel
Mean	1.6	2.9	5	4	18	20	40	13		Gravel

Notes

* The sum of 2 beds separated by 0.5 m of waste
† The sum of 2 beds separated by 0.8 m of waste
‡ The sum of 2 beds separated by 0.6 m of waste
§ Brackets show that the sand and gravel in this borehole does not meet the criteria a and/or b of the definition of mineral (see p. 1); nevertheless, these data have been used in the calculation of the resources.

River Welland and covers 33.8 km², of which 31.0 km² (92.0 per cent) is classified as mineral (all exposed).

Excluding worked ground, the non-mineral areas are confined to isolated patches [005 063, 009 059, 023 063 and 044 075] flanking the River Welland upstream near Stamford. These small areas, investigated either by drilling or field inspection, proved to be mainly sandy clay, usually with locally derived limestone gravel. The composition and surface expression of these deposits is often complicated by local solifluction.

For assessment purposes, the mineral-bearing areas of this block are divided into sub-blocks F_1, F_2 and F_3. Here, the subscript order 1–3 implies a degree of mineral potential of decreasing importance based on mineral area and volume. The sub-blocks are described below.

Sub-block F_1

This, the largest sub-block of block F, extends from near Pilsgate Grange [065 062] in the west to Peakirk in the east, doubling back on itself northwards to incorporate Maxey. It is 19.4 km² in area, of which 18.1 km² is exposed mineral; the remaining 1.3 km² has been worked in two main areas—east of Bainton and west and south of Maxey (Table 3). Both pits are currently expanding.

The assessment is based on 19 IMAU boreholes, one Hydrogeology Unit borehole record and ten others.

The overburden, consisting mainly of soil with clayey gravel ranges from 0.2 to 1.7 m in thickness, these values being recorded from boreholes 00 NE 63 and 10 NW 35—10 SW 10 and 10 NE 26, respectively, with a mean thickness of 0.8 m.

The sub-drift geological lines demonstrate that bedrock changes rapidly across the sub-block from west to east to include seven different solid deposits ranging in age from Upper Estuarine 'Series' to Oxford Clay.

The proven mineral ranges from 0.5 m (borehole 10 SW 10) to 5.5 m (borehole 10 NW 31) in thickness and has a mean of 3.2 m. (In borehole 10 SW 10 the sand and gravel did not meet the criteria a and/or b of the definition of mineral but nevertheless the data have been included in the calculation of the mean thickness

Table 6 Mean composition by weight of gravel (+4–64 mm) in IMAU boreholes in sub-block A_1

Borehole No.	Composition percentage					
	Limestone (including chalk)	Flint	Ironstone	Sandstone	Quartzite	Others
00 NW 56	72	18	10	0	0	0
00 NW 57	25	12	60	1	0	2
00 NW 59	No data available					
00 NW 60	27	38	20	0	15	0
00 NE 43	No data available					
00 NE 44	50	18	24	1	7	0
00 NE 51	45	18	28	0	8	1
00 NE 56	36	19	31	2	12	0
00 NE 61	28	30	31	2	9	0
10 NW 29	59	17	16	1	6	1
10 NW 32	35	12	12	0	40	1
10 NW 33	41	7	17	33	2	0
10 NW 37	48	25	21	2	4	0
10 NW 39	76	11	5	4	4	0
10 NW 40	55	9	18	0	18	0
10 NW 41	58	11	25	1	4	1
10 NW 42	46	23	23	1	6	1
10 NW 45	No data available					
10 NE 29	55	22	8	0	14	1
10 NE 30	27	29	19	1	24	0
10 NE 34	53	14	19	0	14	0
10 NE 35	56	10	22	0	12	0
10 NE 36	47	16	7	23	5	2
10 NE 41	No data available					
10 NE 42	33	39	6	6	10	6
10 NE 44	Non-mineral					
10 NE 48	34	28	16	4	18	0
Mean	45	19	20	4	11	1

of mineral for the estimation of volume). Sand and gravel was absent in only one borehole (00 NE 52).

Seven of the IMAU boreholes proved mineral classified as gravel, another seven proved sandy gravel, two proved 'clayey' sandy gravel, one proved 'very clayey' gravel and one proved 'very clayey' pebbly sand (Table 7). There is an apparent gradation in the deposit from gravel in the west to sandy gravel in the east (Figure 7).

The gravel (+4–64 mm) consists mostly of locally derived oolitic and shelly limestone (45 per cent), flint (24 per cent), ironstone (19 per cent), quartzite (8 per cent) and small amounts of sandstone (3 per cent). (The figures in brackets are mean values.) The three main constituents vary considerably in frequency (limestone from 16 to 60 per cent; flint 11 to 71 per cent; and ironstone 4 to 34 per cent) (Table 8 and Figure 5).

The mean grading for this sub-block is 45 per cent gravel, 48 per cent sand and 7 per cent fines; hence, the mineral in the sub-block is classified overall as sandy gravel. The volume of mineral is estimated at 57.9 million $m^3 \pm 24$ per cent at the 95 per cent probability level.

Sub-block F_2

This occupies a triangular area abutting the Upland of the left bank of the Welland. It is 7.0 km^2 in area, of which 6.0 km^2 is exposed mineral; the remaining 1.0 km^2 has been worked for aggregate north of West Deeping village [110 088] (Table 3). This pit, one of the largest in the area, is continuing to expand.

The assessment is based on eight IMAU boreholes (of which only borehole 00 NE 60 did not prove mineral), two Hydrogeology Unit borehole records and six others.

The overburden (with a recorded thickness of 0.1 to 1.0 m) consists of soil overlying clay with some gravel; the mean thickness is 0.4 m.

The bedrock changes in a north-easterly direction from Kellaways Clay to Oxford Clay, the latter downfaulted at the eastern limit of a postulated north-west to south-east trending fault near Tallington [092 082]. Kellaways Sand was proved along the southern margin of the sub-block and may underlie the eastern extremity.

The proved thickness of mineral ranges from nil in borehole 00 NE 60 to 5.6 m in borehole 00 NE 55 with a mean of 3.5 m. (The 'nil' value has been included in the calculation of the mean thickness of mineral for the estimation of volume.) Generally, the deposits appear to thicken eastwards.

Four of the IMAU boreholes proved mineral classified as gravel, one proved sandy gravel, two proved 'clayey' sandy gravel and one proved waste (Table 9). There is no marked variation of the grading of mineral within the sub-block (Figure 7).

The gravel (+4–64 mm) comprises locally derived oolitic and shelly limestone (58 per cent), flint (19 per cent), ironstone (16 per cent), and quartzite (5 per

Table 7 Sub-block F_1: data from IMAU boreholes

Borehole No.	Recorded thickness		Mean grading percentage						Grading classification
	Overburden m	Mineral m	Fines $-\frac{1}{16}$ mm	Fine sand $+\frac{1}{16}-\frac{1}{4}$ mm	Medium sand $+\frac{1}{4}-1$ mm	Coarse sand $+1-4$ mm	Fine gravel $+4-16$ mm	Coarse gravel $+16$ mm	
00 NE 52	Absent	Absent	—	—	—	—	—	—	
00 NE 57	0.9	5.1	5	6	21	17	38	13	Gravel
00 NE 58	1.3	3.6	5	5	20	16	42	12	Gravel
00 NE 62	1.0	3.4	8	5	20	21	35	11	Sandy gravel
00 NE 63	0.2	4.1	13	13	28	14	28	4	'Clayey' sandy gravel
00 NE 64	1.0	1.6	20	12	17	9	34	8	'Very Clayey' gravel
00 NE 66	1.1	2.9	7	6	23	20	36	8	Sandy gravel
00 SE 40	0.9	3.1	6	7	22	17	36	12	Gravel
10 NW 31	0.8	5.5	6	6	21	19	37	11	Gravel
10 NW 34	0.5	4.8*	4	3	19	21	42	11	Gravel
10 NW 35	0.2	5.0	12	10	25	16	31	6	'Clayey' sandy gravel
10 NW 38	0.4	5.1	4	4	20	19	43	10	Gravel
10 NW 43	1.0	1.8	5	8	33	17	33	4	Sandy gravel
10 NW 46	0.9	3.2	6	3	19	17	46	9	Gravel
10 NW 47	0.5	2.2	9	9	23	20	36	3	Sandy gravel
10 SW 10	1.7	(0.5)†	37	27	21	7	8	0	'Very clayey' pebbly sand
10 NE 26	1.7	3.7	5	7	27	21	32	8	Sandy gravel
10 NE 27	0.9	2.7	8	6	24	22	34	6	Sandy gravel
10 NE 31	1.0	4.1	4	6	25	20	36	9	Sandy gravel
Mean	0.8	3.2	7	7	23	18	36	9	Sandy gravel

Notes
* The sum of 2 beds separated by 0.7 m of waste
† Brackets—see Table 5

Table 8 Mean composition by weight of gravel (+4–64 mm) in IMAU boreholes in sub-block F_1

Borehole No.	Composition percentage					
	Limestone (including chalk)	Flint	Ironstone	Sandstone	Quartzite	Others
00 NE 52	Non-mineral	—	—	—	—	—
00 NE 57	51	18	23	1	5	2
00 NE 58	55	21	13	0	9	2
00 NE 62	42	13	29	1	14	1
00 NE 63	43	14	28	4	11	0
00 NE 64	57	21	9	0	9	4
00 NE 66	42	12	34	0	12	0
00 SE 40	60	22	12	5	1	0
10 NW 31	21	71	7	0	0	1
10 NW 34	55	11	17	14	2	1
10 NW 35	38	16	25	15	6	0
10 NW 38	53	18	25	0	3	1
10 NW 43	46	15	21	3	15	0
10 NW 46	16	69	4	6	5	0
10 NW 47	No data available	—	—	—	—	—
10 SW 10	No data available	—	—	—	—	—
10 NE 26	45	14	21	1	15	4
10 NE 27	33	31	16	1	17	2
10 NE 31	49	14	24	0	11	2
Mean	45	24	19	3	8	1

Table 9 Sub-block F_2: data from IMAU boreholes

Borehole No.	Recorded thickness		Mean grading percentage						Grading classification
	Overburden	Mineral	Fines	Fine sand	Medium sand	Coarse sand	Fine gravel	Coarse gravel	
	m	m	$-\frac{1}{16}$ mm	$+\frac{1}{16}-\frac{1}{4}$ mm	$+\frac{1}{4}-1$ mm	+1–4 mm	+4–16 mm	+16 mm	
00 NE 53	0.5	4.1	4	4	18	17	42	15	Gravel
00 NE 54	0.4	2.7	5	8	21	17	37	12	Gravel
00 NE 55	0.4	5.6	10	8	21	16	32	13	'Clayey' sandy gravel
00 NE 59	0.4	2.2	10	11	32	14	28	5	'Clayey' sandy gravel
00 NE 60	Absent	Absent	—	—	—	—	—	—	—
00 NE 65	0.9	1.3	6	9	28	17	32	8	Sandy gravel
10 NW 28	1.0	5.3	5	6	19	19	12	12	Gravel
10 NW 36	0.9	5.4	4	5	24	16	40	11	Gravel
Mean	0.4	3.5	6	7	22	17	37	11	Gravel

Table 10 Mean composition by weight of gravel (+4–64 mm) in IMAU boreholes in sub-block F_2

Borehole No.	Composition percentage					
	Limestone (including chalk)	Flint	Ironstone	Sandstone	Quartzite	Others
00 NE 53	53	21	19	0	7	0
00 NE 54	58	24	16	1	1	0
00 NE 55	63	12	19	0	5	1
00 NE 59	48	37	12	1	2	0
00 NE 60	Non-mineral	—	—	—	—	—
00 NE 65	56	11	16	0	17	0
10 NW 28	72	17	9	0	2	0
10 NW 36	58	13	18	8	2	1
Mean	58	19	16	1	5	1

cent)—see Table 10 and Figure 5. (The figures quoted in brackets represent mean values for the sub-block as a whole.)

The mean grading of the mineral is gravel 48 per cent, sand 46 per cent, and fines 6 per cent and hence the mineral in the sub-block is classified overall as gravel. The volume is estimated as 21.0 million $m^3 \pm 26$ per cent at the 95 per cent probability level.

Sub-block F_3 This sub-block comprises the remaining areas of First Terrace classified as exposed mineral and is thus represented as isolated areas thoughout the Welland valley, the Gwash valley and in the Fenland.

Some small occurrences included within this sub-block may not individually meet all the criteria laid down (Appendix B, paragraph 12) but the collective area of mineral-bearing ground is sufficient to justify their assessment.

Four of the larger of these spreads occur in the Northborough [153 078]–Peakirk area and one in the upper Welland valley adjacent to Burghley Park. They range in area between 1.0 and 2.5 km^2.

The sub-block is 7.4 km^2 in area, of which 6.9 km^2 is mineral-bearing, the remaining 0.5 km^2 having been worked-out (Table 3). Only the small pit at [050 070] is still active; the remaining areas are left either derelict, used as refuse dumps (for example at [175 082]) or allowed to fill with water for conservation and recreational purposes (for example at [182 080]).

The assessment is based on 13 IMAU boreholes (all of which proved mineral), one Hydrogeology Unit borehole record and six others.

The recorded overburden ranged from 0.3 to 4.5 m in thickness (in boreholes 00 NW 63 and 00 NW 55, respectively) with a mean thickness of 1.0 m (Table 11). The greatest thicknesses occur in the upper reaches of the River Gwash (for example in boreholes 00 NW 55 and 00 NW 61) where the overburden probably includes soliflucted material.

The thickness of mineral proved in IMAU boreholes ranged from 1.6 to 5.4 m (in boreholes 10 NE 45 and 00 NE 45, respectively) with a mean thickness for the sub-block of 3.3 m.

Three boreholes proved mineral classified as gravel, four proved sandy gravel, five proved 'clayey' sandy gravel, and one proved 'very clayey' pebbly sand (Table 11). All five boreholes proving 'clayey' sandy gravel occur in the upper reaches of either the River Welland or the Gwash (Figure 7).

The gravel comprises mostly oolitic and shelly limestone ranging from 14 per cent (in borehole

00 NW 58) to 80 per cent (in borehole 00 NW 61) with a mean of 47 per cent for the sub-block.

Ironstone, the second most common mineral, varies from 8 to 70 per cent (in boreholes 00 NW 61 and 00 NW 58, respectively), with a mean of 28 per cent. The abnormally high ironstone content in 00 NW 58 is probably related to outcrops of Northampton Sand occurring upstream. Flint ranges from 1 to 22 per cent (in boreholes 00 NW 55 and 10 NE 38, respectively), with a mean of 12 per cent. Of the remaining minor constituents only quartzite, particularly in borehole 00 NW 55, was proved in significant quantities (Table 12).

The mean grading of the mineral in this sub-block is

Table 11 Sub-block F_3: data from IMAU boreholes

Borehole No.	Recorded thickness		Mean grading percentage						Grading classification
	Overburden	Mineral	Fines	Fine sand	Medium sand	Coarse sand	Fine gravel	Coarse gravel	
	m	m	$-\frac{1}{16}$ mm	$+\frac{1}{16}-\frac{1}{4}$ mm	$+\frac{1}{4}-1$ mm	$+1-4$ mm	$+4-16$ mm	$+16$ mm	
00 NW 55	4.5	2.6	13	9	15	20	32	11	'Clayey' sandy gravel
00 NW 58	0.4	3.9*	13	18	26	11	23	9	'Clayey' sandy gravel
00 NW 61	2.7	2.2	19	10	19	21	26	5	'Clayey' sandy gravel
00 NW 63	0.3	2.9	15	11	21	16	29	8	'Clayey' sandy gravel
00 NW 64	0.7	5.1†	9	18	29	11	27	6	Sandy gravel
00 NE 45	0.6	5.4	9	9	26	14	30	12	Sandy gravel
00 NE 46	1.3	2.3‡	13	17	36	9	17	8	'Clayey' sandy gravel
10 NW 44	0.6	4.3	4	4	16	19	42	15	Gravel
10 NW 48	1.4	2.5	5	6	22	17	38	12	Gravel
10 NE 25	1.0	1.9	6	10	23	19	35	7	Sandy gravel
10 NE 38	1.6	1.7	5	8	20	21	39	7	Sandy gravel
10 NE 40	1.2	3.3	4	3	16	25	43	9	Gravel
10 NE 45	1.1	1.6	22	20	45	7	5	1	'Very clayey' pebbly sand
Mean	1.0	3.3	11	10	24	16	31	8	'Clayey' sandy gravel

Notes
* the sum of 2 beds separated by 2.1 m of waste
† the sum of 2 beds separated by 1.2 m of waste
‡ the sum of 2 beds separated by 0.4 m of waste

Table 12 Mean composition by weight of gravel (+4–64 mm) in IMAU boreholes in sub-block F_3

Borehole No.	Composition percentage					
	Limestone (including chalk)	Flint	Ironstone	Sandstone	Quartzite	Others
00 NW 55	20	1	29	1	49	0
00 NW 58	14	9	70	0	4	3
00 NW 61	80	10	8	0	2	0
00 NW 63	78	3	11	0	2	6
00 NW 64	63	16	21	0	0	0
00 NE 45	54	14	25	5	2	0
00 NE 46	40	17	40	0	3	0
10 NW 44	49	11	22	6	11	1
10 NW 48	41	17	34	0	6	2
10 NE 25	40	17	24	1	17	1
10 NE 38	35	22	29	0	13	1
10 NE 40	54	12	25	1	8	0
10 NE 45	No data available					
Mean	47	12	28	1	10	2

gravel 39 per cent, sand 50 per cent, and fines 11 per cent and hence it is classified as a 'clayey' sandy gravel. The volume of mineral is estimated to be 22.8 million $m^3 \pm 20$ per cent at the 95 per cent probability level.

Block G
This block includes all Second Terrace deposits which readily subdivide into non-mineral areas (five small areas of terrace flanking the River Welland valley near Stamford) and two mineral-bearing areas (labelled G_1) north-west of Greater Peterborough.

Borehole information and field investigations into the non-mineral terraces at [001 059, 004 064 and 052 078] indicate that the deposits comprise mostly clay with some sand and gravel.

Sub-block G_1 This sub-block is 4.4 km^2 in area, of which just less than 0.2 km^2 is exposed Oxford Clay. The sub-block comprises two unequal areas of exposed mineral; the larger lies at the northern end of an extensive unassessed area of Second Terrace within the built-up area of Greater Peterborough—the other immediately northward of Marholm [147 020]. There are no records of gravel extraction.

The assessment is based on information from six IMAU boreholes, one Hydrogeology Unit borehole record and seven others. Two IMAU boreholes, 10 NE 28 and 10 SE 786, proved sand and gravel less than 1.0 m thick; although these boreholes do not individually meet the criteria a and/or b of the definition of mineral, the data have been used in the calculation of the resources. The overburden consists of soil overlying sandy clay with some gravel and ranged from 0.4 to 1.6 m in thickness (in boreholes 10 SE 786 and 10 NE 28, respectively) with a mean thickness of 0.7 m.

The bedrock is mostly Oxford Clay. However, at the northern end of the smaller sub-block area near Middle Farm [145 032], Kellaways Sand and Kellaways Clay have been mapped.

The proved mineral ranged from 0.4 to 2.4 m in thickness (in boreholes 10 NE 28 and 10 SE 786, respectively) with a mean thickness of 1.3 m.

Of the IMAU boreholes, two proved 'clayey' sandy gravel and the remainder proved sandy gravel, 'clayey' pebbly sand, 'very clayey' pebbly sand and 'clayey' gravel. The results indicate no obvious trends although an anomalous fines content was proved in borehole 10 SE 785 with a mean grading of gravel 5 per cent, sand 56 per cent and fines 39 per cent (Table 13 and Figure 7).

The gravel comprises mainly oolitic and shelly limestone with a mean of 50 per cent. Boreholes 10 SE 786, 10 SW 12 and 10 NE 28 proved relatively high proportions of limestone (with 81, 77 and 45 per cent, respectively); the remaining boreholes in the sub-block showed greater proportions of flint, ironstone and quartzite. Generally, flint is common with a mean of 28 per cent, whereas there are only small amounts of quartzite (12 per cent), ironstone (8 per cent) and sandstone (1 per cent) (Table 14 and Figure 5).

The mean grading of the mineral in the sub-block is gravel 27 per cent, sand 53 per cent and fines 20 per cent; hence, it is classified as 'very clayey' sandy gravel. The volume of mineral is estimated to be 5.5 million $m^3 \pm 39$ per cent) at the 95 per cent probability level.

Block O
This, the largest block, lies entirely within the Upland (pp, 3, 4 and Figures 2 and 8). It comprises three parts: the two smaller parts lie north of the River Welland, and are separated by the River Gwash valley, and the third occupies the southern part of the assessed area south of the River Welland floodplain. The block consists mostly of outcrops of Jurassic rocks ranging in age from Upper Lias to Oxford Clay. Isolated outcrops of Glacial Sand and Gravel, Boulder Clay, Alluvium and River Terrace Deposits have been investigated and *inferred* assessments are offered where mineral is present (p. 1 and Appendix B, paragraph 12).

Boreholes proved a total of 2.2 km^2 (see Table 2b) of mineral in the Glacial Sand and Gravel and beneath the Alluvium and the First Terrace deposits of the River Nene tributary.

The only large area formerly worked for gravel is at [064 088] near Grange Farm Cottages, but there are several small diggings elsewhere, usually for farm track ballast. No records are known to exist regarding other areas of worked ground at [005 045, 017 046, 018 050, 034 060 and 072 051]. Nevertheless, these pits were probably sources of building stone (rather than for sand and gravel) for the local villages, although ironstone from the Northampton Sand was probably extracted in the first two locations as well.

The two mineral-bearing areas within block O, for which an *inferred* assessment is offered, are considered below:

1. *River Terrace Deposits (including sub-Alluvial material)* Boreholes sited in the deposits of the River Nene tributary (unnamed) flowing southwards from Southorpe proved mineral. Conversely, boreholes 00 NW 53 and 00 NW 54 sited in the upper reaches of the River Gwash, north-east of Great Casterton [002 088] proved only soil with clay overlying impersistent thin sandy gravel on bedrock. A small area of First Terrace cropping out at [020 099] near Little Casterton is included in the assessment of sub-block F_3. The area of Second Terrace near Middle Farm [145 032], Marholm, is part of the larger spread north of Greater Peterborough and thus has been included in the assessment of sub-block G_1.

An assessment of mineral within the River Nene tributary valley is offered based on information from three IMAU boreholes and one other. The proved mineral ranges in thickness from 4.0 m in borehole 00 SE 37 to 6.5 m in borehole 00 SE 38 with a mean thickness of 6.0 m.

Near the southern limit of the sheet between Sacrewell Lodge [075 009] and Sutton Heath [090 005], the valley contains First Terrace sand and gravel which carries only thin spreads of soil with clay. Along the valley floor, Alluvium (proved up to 3.6 m thick in borehole 00 SE 2) consisting of clay with some gravel overlies the First Terrace deposit. North of [082 012] no First Terrace is exposed. The mean recorded thickness of overburden is 1.3 m.

Because this First Terrace deposit lies at a higher altitude (16.0 to 18.9 m; 52.5 to 62.0 ft above Ordnance Datum) compared with the River Nene First Terrace deposits (8 to 13 m; 26 to 42 ft above Ordnance Datum) exposed in the valley south of the map, it might be preferable to correlate this deposit with the Third Terrace of the River Nene exposed immediately

Table 13 Sub-block G_1: data from IMAU boreholes

Borehole No.	Recorded thickness		Mean grading percentage						Grading classification
	Overburden	Mineral	Fines	Fine sand	Medium sand	Coarse sand	Fine gravel	Coarse gravel	
	m	m	$-\frac{1}{16}$ mm	$+\frac{1}{16}-\frac{1}{4}$ mm	$+\frac{1}{4}-1$ mm	+1–4 mm	+4–16 mm	+16 mm	
10 NE 28	1.6	(0.4)§	19	26	22	7	21	5	'Clayey' sandy gravel
10 NE 32	0.8	1.5	13	9	18	16	30	14	'Clayey' gravel
10 NE 33	0.8	1.1	5	27	22	7	27	12	Sandy gravel
10 SE 785	0.6	2.4	39	30	23	3	3	2	'Very clayey' pebbly sand
10 SE 786	0.4	(0.8)§	13	16	51	5	12	3	'Clayey' pebbly sand
10 SW 12	1.0	1.6	10	13	17	17	35	8	'Clayey' sandy gravel
Mean	0.7	1.3	20	20	24	9	20	7	'Very clayey' sandy gravel

Note
§Brackets—see Table 5

Table 14 Mean composition by weight of gravel (+4–64 mm) in IMAU boreholes in sub-block G_1

Borehole No.	Composition percentage					
	Limestone (including chalk)	Flint	Ironstone	Sandstone	Quartzite	Others
10 NE 28	45	33	11	1	10	0
10 NE 32	28	40	18	2	12	0
10 NE 33	21	41	8	2	28	0
10 SE 785	No data available	—	—	—	—	—
10 SE 786	81	7	2	0	5	5
10 SW 12	77	17	2	1	2	1
Mean	50	28	8	1	12	1

south of the assessed area (personal communication, A. M. Harrisson). Further data are needed to support this supposition.

The three IMAU boreholes prove 'clayey' gravel, 'clayey' sandy gravel, and gravel. The gravel (+4–64 mm) comprises mainly oolitic and shelly limestone with a mean value of 58 per cent by weight, approximately equal proportions of ironstone and flint (17 and 16 per cent, respectively), 8 per cent quartzite and 1 per cent sandstone (Tables 15 and 16).

The mean grading for the fluvial gravel in the valley at Southorpe is gravel 46 per cent, sand 45 per cent and fines 9 per cent. The *inferred* volume of mineral is estimated to be 5.9 million m^3.

2 *Glacial deposits* Outcrops of Glacial Sand and Gravel and of Boulder Clay are confined to block O in this assessment area. Although mineral was not proved beneath Boulder Clay, it does occur in four of the eight areas of exposed Glacial Sand and Gravel investigated. These deposits occur east and north-east of Stamford at [065 073, 074 097 and 063 088] and 2.5 km west of Greater Peterborough near Belsize Farm [139 011]. In the last locality, the greater part of the mineral has been worked out (Table 3).

The assessment is based on information from six IMAU boreholes, four of which proved mineral. The exceptions, boreholes 00 NE 48 [0756 0925] and 00 NE 49 [0761 0822], proved Glacial Sand and Gravel comprising mostly clay with some limestone gravel around and to the south of Casewick Hall [077 090], Uffington.

Recorded thicknesses of overburden, which generally consists of sandy soil ranged from 0.1 to 1.6 m thick, with a mean thickness of 0.7 m (see Table 1).

The proved mineral ranged from 1.2 to 4.1 m in thickness (in boreholes 00 NE 67 and 00 NE 47, respectively) with a mean thickness of 2.2 m.

Three boreholes proved 'clayey' sandy gravel and one proved 'very clayey' pebbly sand (see Table 15). The composition of the gravel (+4–64 mm) is mainly shelly and oolitic limestone (58 per cent), with flint (26 per cent), ironstone (10 per cent) and the remaining 6 per cent mostly of sandstone (3 per cent) and quartzite (2 per cent) (Table 16). Figures are mean values for the block.

The mean grading of the mineral proved in the Glacial Sand and Gravel in the block is gravel 37 per cent, sand 47 per cent and fines 16 per cent. The *inferred* volume of mineral is estimated to be 2.8 million m^3.

Table 15 Block O: data from IMAU boreholes

Borehole No.	Recorded thickness		Mean grading percentage						Grading classification
	Overburden	Mineral	Fines	Fine sand	Medium sand	Coarse sand	Fine gravel	Coarse gravel	
	m	m	$-\frac{1}{16}$ mm	$+\frac{1}{16}-\frac{1}{4}$ mm	$+\frac{1}{4}-1$ mm	+1–4 mm	+4–16 mm	+16 mm	
1 First Terrace (Nene tributary) partially obscured by Alluvium									
00 SE 37	1.0	4.0	12	14	15	11	33	15	'Clayey' gravel
00 SE 38	0.6	6.5	10	9	21	19	34	7	'Clayey' sandy gravel
00 SE 39	0.5	5.4	6	3	13	29	37	12	Gravel
Mean	1.3	6.0	9	8	17	20	35	11	Gravel
2. Glacial Sand and Gravel									
00 NE 47	0.1	4.1	14	10	22	15	24	15	'Clayey' sandy gravel
00 NE 48	Absent	—	—	—	—	—	—	—	—
00 NE 49	Absent	—	—	—	—	—	—	—	—
00 NE 50	0.1	2.7	16	8	16	19	27	14	'Clayey' sandy gravel
00 NE 67	1.6	1.2	25	54	6	6	7	2	'Very clayey' pebbly sand
00 SW 11	0.5	1.9	14	10	22	11	21	22	"Clayey' sandy gravel
Mean	0.7	2.2	16	15	18	14	22	15	'Clayey' sandy gravel

Table 16 Mean composition by weight of gravel (+4–64 mm) in IMAU boreholes in block O

Borehole No.	Composition percentage					
	Limestone (including chalk)	Flint	Ironstone	Sandstone	Quartzite	Others
1 First Terrace (Nene tributary) partially obscured by Alluvium						
TF 00 SE 37	44	17	23	2	14	0
TF 00 SE 38	63	18	13	0	6	0
TF 00 SE 39	66	13	14	0	5	2
Mean	58	16	17	1	8	0
2 Glacial Sand and Gravel						
TF 00 NE 47	68	20	10	0	2	0
TF 00 NE 48	Non-mineral	—	—	—	—	—
TF 00 NE 49	Non-mineral	—	—	—	—	—
TF 00 NE 50	71	12	14	0	3	0
TF 00 NE 67	No data available	—	—	—	—	—
TF 10 SW 11	36	47	2	10	2	3
Mean	58	26	10	3	2	1

APPENDIX A

FIELD AND LABORATORY PROCEDURES

Trial and error during initial studies of the complex and variable glacial deposits of East Anglia and Essex showed that an absolute minimum of five sample points evenly distributed across the sand and gravel are needed to provide a worthwhile statistical assessment, but that, where possible, there should be not less than ten. Sample points are any points for which adequate information exists about the nature and thickness of the deposit and may include boreholes other than those drilled during the survey and exposures. In particular, the cooperation of sand and gravel operators ensures that boreholes are not drilled where reliable information is already available; although this may be used in the calculations, it is held confidentially by the Institute and cannot be disclosed.

In this report the mineral shown on the 1:25 000 sheet is divided into resource blocks and sub-blocks designated by a letter and subscript numbers, respectively. The block boundaries are determined by geological boundaries and each mineral-bearing formation is separately assessed.

A reconnaissance of the ground is carried out to record any exposures and enquiries are made to ascertain what borehole information is available. Borehole sites are then selected to provide an even pattern of sample points at a density of approximately one per square kilometre. However, because broad trends are independently overlain by smaller scale characteristically random variations, it is unnecessary to adhere to a square grid pattern. Thus such factors as access and the need to minimise disturbance to the land and the public are taken into account in siting the holes; at the same time it is necessary to guard against the possibility that ease of access (that is, the positions of roads and farms) may reflect particular geological conditions, which may bias the drilling results.

The drilling machine employed should be capable of providing a continuous sample representative of all unconsolidated deposits, so that the *in-situ* grading can be determined, if necessary, to a depth of 30 m (100 ft) at a diameter of about 152 mm (6 in), beneath different types of overburden. It should be reliable, quiet, mobile and relatively small (so that it can be moved to sites of difficult access). Shell rigs have proved to be almost ideal: these are described as 'percussion' rigs in the borehole logs—see Appendix D.

The rigs are modified to enable deposits above the water table to be drilled 'dry', instead of with water added to facilitate the drilling, to minimise the amount of material drawn in from outside the limits of the hole. The samples thus obtained are representative of the *in-situ* grading, and satisfy one of the most important aims of the survey. Below the water table the rigs are used conventionally although this may result in the loss of some of the fines fraction and the pumping action of the bailer tends to draw unwanted material into the hole from the sides or the bottom.

A continuous series of bulk samples is taken throughout the sand and gravel. Ideally samples are composed exclusively of the whole of the material encountered in the borehole between the stated depths. However, care is taken to discard, as far as possible, material which has caved or which has been pumped from the bottom of the hole. A new sample is commenced whenever there is an appreciable lithological change within the sand and gravel, or at every 1 m (3.3 ft) depth. The samples, each weighing between 25 and 45 kg (55 and 100 lb), are despatched in heavy duty polythene bags to a laboratory for grading. The grading procedure is based on British Standard 1377 (1967). Random checks on the accuracy of the grading are made in the Institute's laboratories.

In this survey further exploratory drilling was undertaken using a Minuteman power auger rig. This machine, which is small and portable (it will fit into the rear of a long wheel-base Landrover) may be operated by one person; its use is restricted to those occasions when access to land is not possible with shell rigs or when information is required quickly.

The auger tool comprises a continuous 'flight' 76 mm (3 in) auger; the use of this equipment, in addition to 'open hole' drilling methods, inevitably results in mixing and contamination of the sampled material. Thus, data relating to depth and particularly composition cannot be as accurately determined as with shell rigs. Therefore, in this report the Minuteman borehole logs do not show grading data; composition data when present are intended as a guide only.

All data, including mean grading analysis figures calculated for the total thickness of the mineral, are entered on standard record sheets, abbreviated copies of which are reproduced in Appendix F.

Detailed records may be consulted at the appropriate offices of the Institute, upon application to the Head, Industrial Minerals Assessment Unit.

APPENDIX B

STATISTICAL PROCEDURE

Statistical assessment

1 A statistical assessment is made of an area of mineral greater than 2 km^2, if there is a minimum of five evenly spaced boreholes in the resource block (for smaller areas see paragraph 12 below).

2 The simple methods used in the calculations are consistent with the amount of data provided by the survey. Conventional symmetrical confidence limits are calculated for the 95 per cent probability level, that is, there is a 5 per cent or one in twenty chance of a result falling outside the stated limits.

3 The volume estimate (V) for the mineral in a given block is the product of the two variables, the sampled areas (A) and the mean thickness ($\bar{l}_m$) calculated from the individual thicknesses at the sample points. The standard deviations for these variables are related such that

$$S_V = \sqrt{(S_A{}^2 + S_{\bar{l}_m}{}^2)} \quad . \qquad [1]$$

4. The above relationship may be transposed such that

$$S_V = S_{\bar{l}_m}\sqrt{(1 + S_A{}^2/S_{\bar{l}_m}{}^2)} \quad . \qquad [2]$$

From this it can be seen that as $S_A{}^2/S_{\bar{l}_m}{}^2$ tends to 0, S_V tends to $S_{\bar{l}_m}$.

If, therefore, the standard deviation for area is small with respect to that for mean thickness, the standard deviation for volume approximates to that for mean thickness.

5 Given that the number of approximately evenly spaced sample points in the sampled area is n with mineral thickness measurements $l_{m1}, l_{m2}, \ldots, l_{m_n}$, then the best estimate of mean thickness, $\bar{l}_m$, is given by

$$\sum (l_{m1} + l_{m2} \ldots l_{m_n})/n \quad .$$

For groups of closely spaced boreholes a discretionary weighting factor may be applied to avoid bias (see note on weighting below). The standard deviation for mean thickness $S_{\bar{l}_m}$, expressed as a proportion of the mean thickness, is given by

$$S_{\bar{l}_m} = (1/\bar{l}_m)\sqrt{[\sum (l_m - \bar{l}_m)^2/(n-1)]} \ ,$$

where l_m is any value in the series l_{m1} to l_{m_n}.

6 The sampled area in each resource block is coloured pink on the map. Wherever possible, calculations relate to the mineral within mapped geological boundaries (which may not necessarily correspond to the limits of deposits). Where the area is not defined by a mapped boundary, that is, where the boundary is inferred, a distinctive symbol is used. Experience suggests that the errors in determining area are

small relative to those in thickness. The relationship $S_A/S_{\bar{l}_m} \leqq \frac{1}{3}$ is assumed in all cases. It follows from equation [2] that

$$S_{\bar{l}_m} \leqq S_V \leqq 1.05\, S_{\bar{l}_m} \qquad [3]$$

7 The limits on the estimate of mean thickness of mineral, $L_{\bar{l}_m}$, may be expressed in absolute units $\pm(t/\sqrt{n}) \times S_{\bar{l}_m}$ or as a percentage $\pm(t/\sqrt{n}) \times S_{\bar{l}_m} \times (100/\bar{l}_m)$ per cent, where t is Student's t at the 95 per cent probability level for $(n-1)$ degrees of freedom, evaluated by reference to statistical tables. (In applying Student's t it is assumed that the measurements are distributed normally.)

8 Values of t at the 95 per cent probability level for values of n up to 20 are as follows:

n	t	n	t
1	infinity	11	2.228
2	12.706	12	2.201
3	4.303	13	2.179
4	3.182	14	2.160
5	2.776	15	2.145
6	2.571	16	2.131
7	2.447	17	2.120
8	2.365	18	2.110
9	2.306	19	2.101
10	2.262	20	2.093

(from table 12, Biometrika Tables for Statisticians, Volume 1, Second Edition, Cambridge University Press, 1962). When n is greater than 20, 1.96 is used (the value of t when n is infinity).

9 In calculating confidence limits for volume, L_V, the following inequality corresponding to equation [3] is applied: $L_{\bar{l}_m} \leqq L_V \leqq 1.05 L_{\bar{l}_m}$.

10 In summary, for values of n between 5 and 20, L_V is calculated as

$$[(1.05 \times t)/\bar{l}_m] \times [\sqrt{\textstyle\sum}(l_m-\bar{l}_m)^2/n(n-1)] \times 100$$

per cent, and when n is greater than 20, as

$$[(1.05 \times 1.96)/\bar{l}_m] \times [\sqrt{(\textstyle\sum}(l_m-\bar{l}_m)^2/n(n-1)] \times 100$$

per cent (weighting factors may be included: see paragraph 15).

11 The application of this procedure to a fictitious area is illustrated in Figures 9 and 10.

Inferred assessment

12 If the sampled area of mineral in a resource block is between 0.25 km^2 and 2 km^2 an assessment is inferred, based on geological and topographical information usually supported by the data from one or two boreholes. The volume of mineral is calculated as the product of the area, measured from field data, and the estimated thickness. Confidence limits are not calculated.

13 In some cases a resource block may include an area left uncoloured on the map, within which mineral (as defined) is interpreted to be generally absent. If there is reason to believe that some mineral may be present, an inferred assessment may be made.

14 No assessment is attempted for an isolated area of mineral less than 0.25 km^2.

15 *Note on weighting* The thickness of a deposit at any point may be governed solely by the position of the point in relation to a broad trend. However, most sand and gravel deposits also exhibit a random pattern of local, and sometimes considerable, variation in thickness. Thus the distribution of sample points need be only approximately regular and in estimating the mean thickness only simple weighting is necessary. In practice, equal weighting can often be applied to thicknesses at all sample points. If, however, there is a distinctly unequal distribution of points, bias is avoided by dividing the sampled area into broad zones, to each of which a value roughly proportional to its area is assigned. This value is then shared between the data points within the zone as the weighting factor.

APPENDIX C

CLASSIFICATION AND DESCRIPTION OF SAND AND GRAVEL

For the purposes of assessing resources of sand and gravel a classification should take account of economically important characteristics of the deposit, in particular the absolute content of fines and the ratio of sand to gravel.

The terminology commonly used by geologists when describing sedimentary rocks (Wentworth, 1922) is not entirely satisfactory for this purpose. For example, Wentworth proposed that a deposit should be described as a 'gravelly sand' when it contains more sand than gravel and there is at least 10 per cent of gravel, provided that there is less than 10 per cent of material finer than sand (less than $\frac{1}{16}$ mm) and coarser than pebbles (more than 64 mm in diameter). Because deposits containing more than 10 per cent fines are not embraced by this system a modified binary classification based on Willman (1942) has been adopted.

When the fines content exceeds 40 per cent the material is not considered to be potentially workable and falls outside the definition of mineral. Deposits which contain 40 per cent fines or less are classified primarily on the ratio of sand to gravel but qualified in the light of the fines content, as follows: less than 10 per cent fines—no qualification; 10 per cent or more but less than 20 per cent fines—'clayey'; 20 to 40 per cent fines—'very clayey'.

The term 'clay' (as written, with single quote marks) is used to describe all material passing $\frac{1}{16}$ mm. Thus it has no mineralogical significance and includes particles falling within the size range of silt. The normal meaning applies to the term clay where it does not appear in single quotation marks.

The ratio of sand to gravel defines the boundaries between sand, pebbly sand, sandy gravel and gravel (at 19:1, 3:1 and 1:1).

Thus it is possible to classify the mineral into one of twelve descriptive categories (see Figure 11). The procedure is as follows:

1 Classify according to ratio of sand to gravel.
2 Describe fines.

For example, a deposit grading 11 per cent gravel, 70 per cent sand and 19 per cent fines is classified as 'clayey' pebbly sand. This short description is included in the borehole log (see Note 11, Appendix D).

Many differing proposals exist for the classification of the grain size of sediments (Atterberg, 1905; Udden, 1914; Wentworth, 1922; Wentworth, 1935; Allen, 1936; Twenhofel, 1937; Lane and others, 1947). As Archer (1970a, b) has emphasised, there is a pressing need for a simple metric scale acceptable to both scientific and engineering interests, for which the class limit sizes correspond closely with certain marked changes in the natural properties of mineral particles. For example, there is an important change in the degree of cohesion between particles at about the $\frac{1}{16}$-mm size, which approximates to the generally accepted boundary between silt and sand. These and other requirements are met by a system based on Udden's geometric scale and a simplified form of Wentworth's terminology (Table 17), which is used in this Report.

The fairly wide intervals in the scale are consistent with the general level of accuracy of the qualitative assessments of the resource blocks. Three sizes of sand are recognised, fine $(+\frac{1}{16}-\frac{1}{4}$ mm), medium $(+\frac{1}{4}-1$ mm) and coarse (+1–4 mm). The boundary at 16 mm distinguishes a range of finer gravel (+4–16 mm), often characterised by abundance of worn tough pebbles of vein quartz, from larger pebbles often of notably different materials. The boundary at 64 mm distinguishes pebbles from cobbles. The term 'gravel' is used loosely to denote both pebble-sized and cobble-sized material.

The size distribution of borehole samples is determined by sieve analysis, which is presented by the laboratory as logarithmic cumulative curves (see, for example, British Standard 1377: 1967). In this report the grading is tabulated on the borehole record sheets (Appendix F), the intercepts corresponding with the simple geometric scale $\frac{1}{16}$ mm, $\frac{1}{4}$ mm, 1 mm, 4 mm, 16 mm and so on as required. Original sample grading curves are available for reference at the appropriate office of the Institute.

Each bulk sample is described, subjectively, by a geologist at the borehole site. Subsequently, the descriptive categories of the mineral in each borehole are modified, where necessary, according to results obtained from the mean particle size analyses of these samples.

The relative proportions of the rock types present in the gravel fraction are indicated by the use of the words 'and' or 'with'. For example, 'flint and quartz' indicates very approximate equal proportions with neither constituent accounting for less than about 25 per cent of the whole; 'flint with quartz' indicates that flint is dominant and quartz, the principal accessory rock type, comprises 5 to 25 per cent of the whole. Where the accessory material accounts for less than 5 per cent of the whole, but is still readily apparent, the phrase 'with some' has been used. Rare constituents are referred to as 'trace'.

The terms used in the field to describe the degree of rounding of particles, which is concerned with the sharpness of the edges and corners of a clastic fragment and not the shape (after Pettijohn, 1957), are as follows.

Angular: showing little or no evidence of wear; sharp edges and corners.

Subangular: showing definite effects of wear. Fragments still have their original form but edges and corners begin to be rounded off.

Subrounded: showing considerable wear. The edges and corners are rounded off to smooth curves. Original grain shape is still distinct.

Rounded: original faces almost completely destroyed, but some comparatively flat surfaces may still remain. All original edges and corners have been smoothed off to rather broad curves. Original shape is still apparent.

Well-rounded: no original faces, edges or corners left. The entire surface consists of broad curves; flat areas are absent. The original shape is suggested by the present form of the grain.

Table 17 Classification of gravel, sand and fines

Size limits	Grain size description	Qualification	Primary classification
	Cobble		Gravel
64 mm			
	Pebble	Coarse	
16 mm			
		Fine	
4 mm			
	Sand	Coarse	Sand
1 mm			
		Medium	
$\frac{1}{4}$ mm			
		Fine	
$\frac{1}{16}$ mm			
	Fines (silt and clay)		Fines

Block calculation 1:25 000 Block } Fictitious

Area
Block: 11.08 km²
Mineral: 8.32 km²

Mean thickness
Overburden: 2.5 m
Mineral: 6.5 m

Volume
Overburden: 21 million m³
Mineral: 54 million m³

Confidence limits of the estimate of mineral volume at the 95 per cent probability level: ±20 per cent

That is, the volume of mineral (with 95 per cent probability): 54±11 million m³

Thickness estimate (measurements in metres)

l_o = overburden thickness l_m = mineral thickness

Sample point	Weighting w	Overburden l_o	wl_o	Mineral l_m	wl_m	Remarks
SE 14	1	1.5	1.5	9.4	9.4	IMAU boreholes
SE 18	1	3.3	3.3	5.8	5.8	
SE 20	1	nil	—	6.9	6.9	
SE 22	1	0.7	0.7	6.4	6.4	
SE 23	1	6.2	6.2	4.1	4.1	
SE 24	1	4.3	4.3	6.4	6.4	
SE 17	$\frac{1}{2}$	1.2	1.6	9.8	7.2	
123/45	$\frac{1}{2}$	2.0		4.6		Hydrogeology Unit record
1	$\frac{1}{4}$	2.7	2.6	7.3	5.8	Close group of four boreholes (commercial)
2	$\frac{1}{4}$	4.5		3.2		
3	$\frac{1}{4}$	0.4		6.8		
4	$\frac{1}{4}$	2.8		5.9		
Totals	$\sum w=8$	$\sum wl_o=20.2$		$\sum wl_m=52.0$		
Means		$\overline{wl_o}=2.5$		$\overline{wl_m}=6.5$		

Calculation of confidence limits

wl_m	$\lvert(wl_m-\overline{wl_m})$	$(wl_m-\overline{wl_m})^2$
9.4	2.9	8.41
5.8	0.7	0.49
6.9	0.4	0.16
6.4	0.1	0.01
4.1	2.4	5.76
6.4	0.1	0.01
7.2	0.7	0.49
5.8	0.7	0.49

$\sum(wl_m-\overline{wl_m})^2=15.82$
$n=8$
$t=2.365$

L_V is calculated as

$t/w\bar{L}_m$:

$1.05(t/\overline{wl_m})\sqrt{[\sum wl_m-\overline{wl_m})^2/n(n-1)]}+100$
$=1.05\times(2.365/6.5)\sqrt{[15.82/(8\times 7)]}\times 100$
$=20.3$
$\simeq 20$ per cent.

Figure 9 Example of resource block assessment: calculations and results

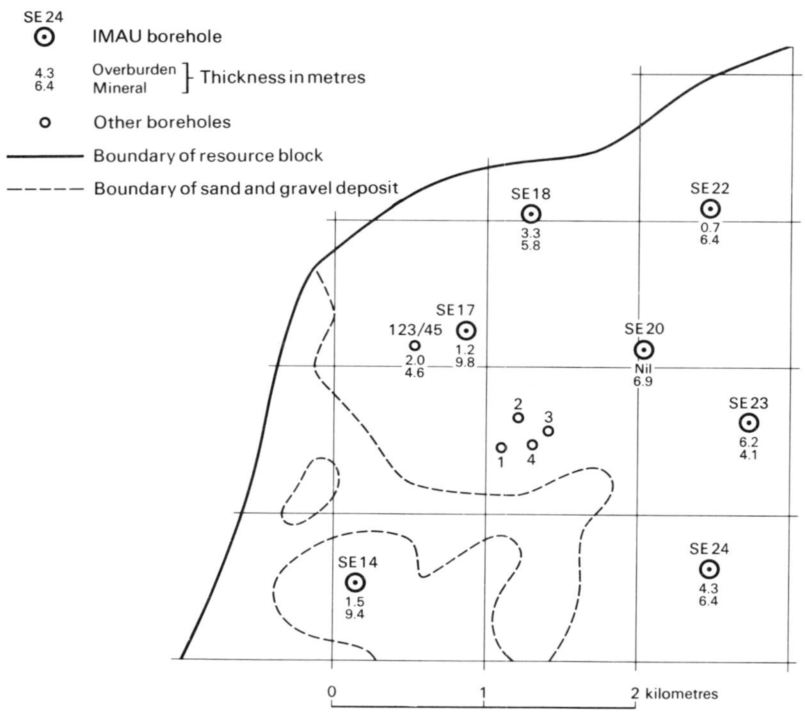

Figure 10 Example of resource block assessment: map of a fictitious block

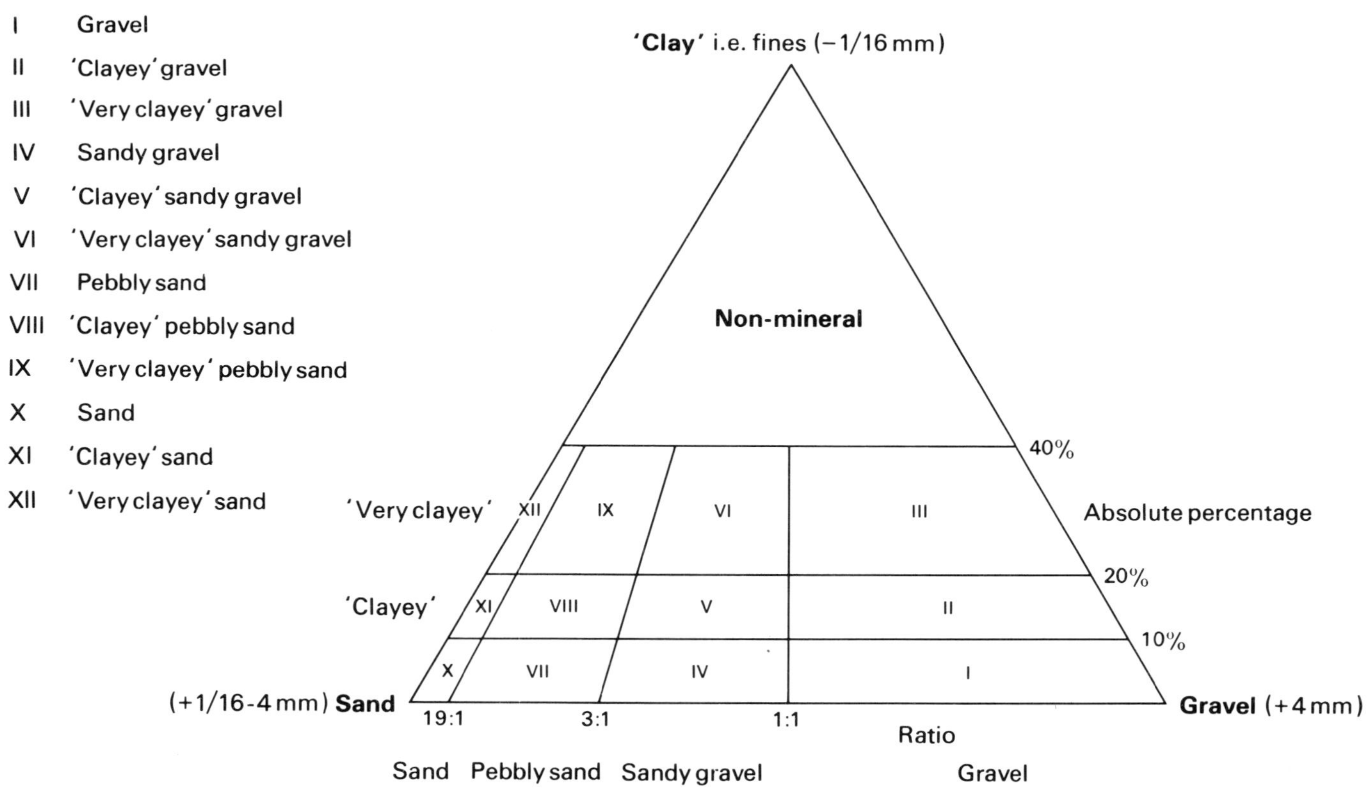

Figure 11 Diagram showing the descriptive categories used in the classification of sand and gravel

APPENDIX D

EXPLANATION OF THE BOREHOLE RECORDS

ANNOTATED EXAMPLE

TF 00 NW 58[1] **0161 0564**[2] **Near Dottrell Hill Plantation, Stamford**[3] **Sub-block F_3**

Surface level (+25.6 m) +84 ft[4]
Water struck at (+22.7 m and +20.7 m)[5]
152 mm percussion[6]
March 1976[6]

[7]Overburden 0.4 m
Mineral 2.4 m
Waste 2.1 m
Mineral 1.5 m
Waste 1.1 m
Bedrock 2.4 m+[9]

LOG

Geological classification[10]	Lithology[11]	Thickness m	Depth m
	Soil, pale reddish brown, with an iron-cemented sand and gravel layer at base	0.4	0.4
River Terrace Deposits (First Terrace)	**a** 'Clayey' sandy gravel Gravel: fine with coarse, ironstone with limestone Sand: medium and fine with trace coarse, subangular to subrounded quartz Fines: orange-brown	2.4	2.8
	Clay, firm to stiff, brown to dark grey, with trace pebbles and shell fragments. Sandy in upper 1.7 m	2.1	4.9
	b Gravel Gravel: fine with some coarse, subrounded to subangular oolitic and shelly limestone with ironstone, flint and quartzite Sand: medium and coarse with trace fine, rounded to subrounded, quartz with ironstone and limestone ooliths Fines: pale greyish brown	1.5	6.4
	Silt, firm to stiff, dark greyish blue, with carbonaceous patches, pebbly towards base	1.1	7.5
Upper Lias	Clay, stiff, dark grey to black, silty, occasional indurated siltstone pellets, trace shell fragments, shaly towards base	2.4+[9]	9.9

GRADING[13]

	Mean for deposit *percentages*			[12]Depth below surface (m)	*percentages*						
	Fines	Sand	Gravel		Fines	Sand			Gravel		
					$-\frac{1}{16}$	$+\frac{1}{16}-\frac{1}{4}$	$+\frac{1}{4}-1$	+1–4	+4–16	+16–64	+64
a	19	60	21	0.4–1.8	17	23	26	6	19	9	0
				1.8–2.8	20	33	31	2	9	5	0
				[14]Mean	19	28	28	4	14	7	0
				2.8–4.9 waste	46	23	23	5	2	1	0
				4.9–5.9	4	4	24	19	37	13	0
				5.9–6.4	4	4	22	23	38	9	0
b	4	48	48	Mean	4	4	23	21	37	11	0
a+b	13	55	32	Mean	13	18	26	11	23	9	0

COMPOSITION[15] (a+b)

Depth below surface (m)	Percentage by weight in gravel (+4–64 mm) fraction					
	Limestone including chalk	Flint	Ironstone	Sandstone	Quartzite	Others
0.4–6.4	14	9	70	0	4	3

The numbered paragraphs below correspond with the annotations given on the specimen record above.

1 Borehole registration number
Each Industrial Minerals Assessment Unit (IMAU) borehole is identified by a registration number. This consists of two statements:

1 The number of the 1:25 000 sheet on which the borehole lies, for example TF 00.
2 The quarter of the 1:25 000 sheet on which the borehole lies and the number of the borehole in a series for that quarter, for example NW 58.

Thus the full registration number is TF 00 NW 58. Usually this is abbreviated to 00 NW 58 in the text.

2 The National Grid reference
All National Grid references in this publication lie within the 100 km square TF unless otherwise stated. Grid references are given to eight figures, accurate to within 10 m for borehole locations. (In the text, six-figure grid references are used for more approximate locations, for example, for farms).

3 Location
The position of the borehole is generally referred to the nearest named locality on the 1:25 000 map, followed by the resource block or sub-block in which it lies.

4 Surface level
The surface level at the borehole site is given in metres and feet above Ordnance Datum. All height measurements were made in feet; their approximate conversions to metres above Ordnance Datum are given in brackets.

5 Groundwater conditions
Four kinds of entry are made: the record indicates the level at which the groundwater was struck (in metres above or below Ordnance Datum); or that water was encountered but its level not recorded; or that water was not struck; or that no note of groundwater conditions was made.

6 Type of drill and date of drilling
Modified shell rigs (percussion) and a portable Minuteman drill were used. The external diameter of the casing used, the type of machine, and the month and year when the borehole was completed are stated.

7 Overburden, mineral, waste and bedrock
Mineral is sand and gravel, which, as part of a deposit, falls within the arbitrary definition of potentially workable material (see p. 1). Bedrock is the 'formation', 'country rock' or 'rock head' below which potentially workable sand and gravel will not be found. Waste is any material other than bedrock or mineral. Where waste occurs between the surface and mineral it is classified as overburden.

8 Thickness and depth
All drilling measurements were made in metres. A conversion table for metres to feet is given in Appendix H.

9 The plus sign (+) indicates that the base of the deposit was not reached during drilling.

10 Geological Classification
The geological classification is given wherever possible.

11 Lithological description
When sand and gravel is recorded a general description based on the mean grading characteristics (for details see Appendix C) is followed by more detailed particulars of the gravel and/or sand and silt fractions. Where more than one mineral stratum is recognised, each is designated by a letter, for example, a, b, etc. The description of other deposits is based on visual examination in the field and, in some instances, laboratory inspection of special samples.

12 Sampling
A continuous series of bulk samples is taken throughout the thickness of sand and gravel. A new sample is commenced whenever there is an appreciable lithological change within the sand and gravel or at every 1 m (3.3 ft) of depth.

13 Grading results
For each bulk sample the percentage of fines ($-\frac{1}{16}$ mm), fine sand ($+\frac{1}{16}$ mm–$\frac{1}{4}$ mm), medium sand ($+\frac{1}{4}$–1 mm), coarse sand (+1–4 mm), fine gravel (+4–16 mm), coarse gravel (+16–64 mm) and cobble gravel (+64 mm) are stated.

14 Mean grading
In calculating mean gradings, data for individual samples are weighted by the thickness represented; the weighted mean grading of groups of samples making up an identified mineral deposit are given in detail and in summary.

Where one or more distinct units of mineral form a continuous sequence separated from another sequence by waste, each mineral unit is designated by a letter (for example a, b, etc.); the weighted mean grading of each is given in detail and in summary; and, the weighted mean grading for the whole borehole (excluding waste sequences) is also given in similar detail.

Fully representative sampling of sand and gravel is difficult to achieve once drilling proceeds beyond the water table when the bailing technique is employed. Comparison between boreholes and adjacent exposures suggests that in bailed borehole samples the proportion of sand may be higher and the proportion of fines and coarse gravel (+16 mm) may be lower.

15 Composition of the gravel (+4–64 mm)
Details of the gravel composition of bulked samples from selected boreholes when available are given. Where more than one stratum is recognised (for example a, b, etc.) the composition proportions for the whole of the mineral in the borehole is given (that is a+b, etc.). However, the percentages shown are intended as a guide only and in this instance do not include any weighting factor (Appendix B, paragraph 15).

APPENDIX E

LIST OF BOREHOLES USED IN THE ASSESSMENT OF RESOURCES

Borehole*	Grid reference†	Block/ sub-block
1 INDUSTRIAL MINERALS ASSESSMENT UNIT BOREHOLES		
TF 00 NW		
53	0033 0882	O
54	0073 0961	O
55	0211 0998	F_3
56	0007 0550	A_1
57	0113 0607	A_1
58	0161 0564	F_3
59	0213 0599	A_1
60	0276 0691	A_1
61	0350 0991	F_3
62	0456 0767	F
63	0493 0773	F_3
64	0438 0711	F_3
66	0029 0641	G
67	0039 0621	F
68	0074 0585	F
69	0027 0637	G
TF 00 NE		
43	0514 0717	A_1
44	0506 0728	A_1
45	0542 0675	F_3
46	0592 0635	F_3
47	0648 0727	O
48	0756 0925	O
49	0761 0822	O
50	0732 0743	O
51	0702 0679	A_1
52	0765 0618	F_1
53	0837 0975	F_2
54	0845 0886	F_2
55	0828 0775	F_2
56	0811 0707	A_1
57	0842 0653	F_1
58	0837 0551	F_1
59	0924 0931	F_2
60	0896 0824	F_2
61	0949 0768	A_1
62	0945 0718	F_1
63	0955 0645	F_1
64	0913 0587	F_1
65	0984 0812	F_2
66	0981 0725	F_1
67	0740 0984	O
TF 00 SE		
37	0791 0188	O
38	0795 0105	O
39	0850 0024	O
40	0921 0494	F_1
TF 10 NW		
28	1097 0952	F_2
29	1016 0860	A_1
31	1035 0630	F_1
32	1164 0889	A_1
33	1105 0830	A_1
34	1191 0798	F_1
35	1116 0654	F_1
36	1159 0942	F_2
37	1275 0883	A_1
38	1216 0731	F_1
39	1319 0992	A_1
40	1346 0912	A_1
41	1305 0793	A_1
42	1317 0696	A_1
43	1310 0583	F_1
44	1402 0863	F_3
45	1397 0785	A_1
46	1406 0674	F_1
47	1396 0589	F_1
48	1488 0912	F_3
TF 10 NE		
25	1509 0870	F_3
26	1545 0768	F_1
27	1502 0661	F_1
28	1542 0539	G_1
29	1631 0979	A_1
30	1626 0886	A_1
31	1645 0787	F_1
32	1627 0646	G_1
33	1630 0540	G_1
34	1768 0854	A_1
35	1724 0757	A_1
36	1742 0672	A_1
37	1752 0582	F
38	1819 0946	F_3
39	1876 0922	D_1
40	1858 0837	F_3
41	1867 0731	A_1
42	1850 0617	A_1
43	1856 0542	D
44	1966 0814	A_1
45	1963 0722	F_3
46	1953 0635	D
47	1966 0544	D
48	1718 0955	A_1
49	1771 0537	F
TF 10 SE		
785	1574 0432	G_1
786	1522 0290	G_1
788	1799 0463	F
790	1823 0422	F
791	1884 0363	D
792	1963 0452	D
793	1942 0357	F
818	1832 0478	F
TF 10 SW		
10	1390 0481	F_1
11	1371 0140	O
12	1497 0329	G_1

Hydrogeology Unit borehole	Borehole No.*‡	Grid reference†	Block/ sub-block
2 OTHER IGS REGISTERED BOREHOLES			
157/8	not registered by 17.01.79	0916 0618	F_1
157/45		1096 0866	F_2
157/73		0934 0786	F_2
157/113		1128 0872	A_1
157/165		1090 0800	A_1
	TF 10 NE		
158/4	17	1548 0596	G_1
159/39	6	1518 0954	F_3
158/105	4	1594 0940	A_1
	TF 00 NW		
	43	0357 0670	O
	44	0366 0676	O
	46	0375 0694	F_3
	48	0377 0707	A_1
	50	0376 0714	A_1
	TF 00 NE		
	41A	0977 0873	F_2
	TF 00 SE		
	2	0841 0205	O
	TF 10 NW		
	1A	1416 0958	A_1
	2	1387 0995	F_3
	3	1396 0650	F_1
	4	1398 0650	F_1
	5	1342 0534	F_1
	6	1424 0508	F_1
	7	1415 0509	F_1
	8	1423 0517	F_1
	9	1287 0562	F_1
	10	1288 0556	F_1
	14	1479 0961	A_1
	TF 10 NE		
	1	1502 0526	G_1
	2	1792 0996	A_1
	3	1589 0972	F_3
	7	1926 0867	A_1
	8A	1914 0848	D_1
	10	1950 0826	A_1
	11	1822 0882	F_3
	13	1743 0815	F_3
	15	1675 0672	G_1
	16	1691 0665	G_1
	18	1636 0510	G_1
	19	1625 0501	G_1
	20	1887 0503	D
	21	1638 0500	G_1
	TF 10 SL		
	176	1624 0497	G_1
	TF 10 SW		
	6	1412 0497	F_1
	7	1408 0493	F_1

CONFIDENTIAL RECORDS

Many records made available by industry are held on a 'commercial-in-confidence' basis

* By sheet quadrant
† All fall within 100 km squares TF 00 and TF 10
‡ See note on 'other boreholes' on the map (in pocket)

APPENDIX F

INDUSTRIAL MINERALS ASSESSMENT UNIT BOREHOLE RECORDS

TF 00 NW 53 0033 0882 Great Casterton, Stamford **Block O**

Surface level (+37.8 m)+124 ft
Water struck at (+33.4 m)
152 mm percussion
March 1976

Waste 5.2 m
Bedrock 2.0 m+

LOG

Geological classification	Lithology	Thickness m	Depth m
	Soil, dark brown	0.2	0.2
Alluvium	Clay, firm, pale brown, sandy, becomes dark grey with shell fragments towards base	4.2	4.4
River Terrace Deposits (First Terrace)	'Clayey' sandy gravel Gravel: fine with trace coarse, predominantly angular to subangular ironstoned oolitic limestone with traces of ironstone, flint and sandstone Sand: medium with coarse and fine, limestone ooliths with traces of shell fragments Fines: dark grey	0.8	5.2
Lower Lincolnshire Limestone	Clay, firm to stiff, dark grey, trace shell fragments	2.0+	7.2

GRADING

Mean for deposit *percentages*			Depth below surface (m)	*percentages*						
Fines	Sand	Gravel		Fines	Sand			Gravel		
				$-\frac{1}{16}$	$+\frac{1}{16}-\frac{1}{4}$	$+\frac{1}{4}-1$	+1–4	+4–16	+16–64	+64
17	56	27	4.4–5.2	17	11	28	27	21	6	0

COMPOSITION

Depth below surface (m)	Percentage by weight in gravel (+4–64 mm) fraction					
	Limestone including chalk	Flint	Ironstone	Sandstone	Quartzite	Others
4.4–5.2	93	2	4	1	0	0

TF 00 NW 54 0073 0961 Glebe Barn, near Great Casterton **Block O**

Surface level (+35.7 m)+117 ft
Waterstruck at (+32.7 m)
152 mm percussion
March 1976

Waste 3.2 m
Bedrock 2.1 m+

LOG

Geological classification	Lithology	Thickness m	Depth m
	Soil, brown	0.2	0.2
Alluvium	Clay, soft to firm, brown to dark grey mottled in places, silty, glutinous in places, with black carbonaceous material and patches of shell fragments below 2.0 m	3.0	3.2
?Northampton Sand	Clay, oolitic sand in upper 0.2 m. Firm to stiff, dark grey to greyish blue, silty, slightly micaceous, occasional shell fragments	2.1+	5.3

TF 00 NW 55 0211 0998 Near Little Casterton, Stamford **Sub-block F_3**

Surface level (+38.7 m)+127 ft
Water not struck
152 mm percussion
March 1976

Overburden 4.5 m
Mineral 2.6 m
Bedrock 2.5 m+

LOG

Geological classification	Lithology	Thickness m	Depth m
	Soil, clayey with limestone fragments	0.3	0.3
?Head/River Terrace Deposits (First Terrace)	Clay, firm, orange-brown becoming khaki below 4.0 m, silty in upper parts, ferruginous with gravel patches and shell fragments	4.2	4.5
River Terrace Deposits	'Clayey' sandy gravel, particularly sandy in upper 0.5 m Gravel: fine with some coarse, mainly quartzite with ironstone and limestone, traces of flint and sandstone Sand: coarse and medium with some fine, quartz with limestone ooliths	2.6	7.1
? Lower Lincolnshire Limestone	Sand, greyish brown becoming reddish brown below 8.2 m, iron-oxide coated quartz in part cemented layers with intercalations of clay	2.5+	9.6

GRADING

Mean for deposit percentages			Depth below surface (m)	percentages						
				Fines	Sand			Gravel		
Fines	Sand	Gravel		$-\frac{1}{16}$	$+\frac{1}{16}-\frac{1}{4}$	$+\frac{1}{4}-1$	+1–4	+4–16	+16–64	+64
13	44	43	4.5–5.5	14	12	20	16	28	10	0
			5.5–6.5	12	7	12	23	36	10	0
			6.5–7.1	14	7	11	20	36	12	0
			Mean	13	9	15	20	32	11	0

COMPOSITION

Depth below surface (m)	Percentage by weight in gravel (+4–64 mm) fraction					
	Limestone including chalk	Flint	Ironstone	Sandstone	Quartzite	Others
4.5–7.1	20	1	29	1	49	0

TF 00 NW 56 0007 0550 Welland valley, near Tinwell, Stamford **Sub-block A_1**

Surface level (+25.7 m) +84.5 ft
Water struck at (+22.4 m)
152 mm percussion
March 1976

Overburden 3.3 m
Mineral 3.6 m
Bedrock 2.1 m+

LOG

Geological classification	Lithology	Thickness m	Depth m
	Soil, pale brown	0.4	0.4
Alluvium	Clay, firm to stiff, pale grey mottled brown and orange-brown, silty	2.9	3.3
River Terrace Deposits (First Terrace)	Gravel Gravel: fine with coarse, predominantly subrounded, tabular, shelly and oolitic limestone with angular to subangular flint with rounded to subrounded ironstone Sand: coarse and medium with trace fine, quartz and ironstone with some limestone ooliths Fines: brown to pale brown	3.6	6.9
Upper Lias	Clay, stiff, dark grey, silty	2.1+	9.0

GRADING

Mean for deposit *percentages*			Depth below surface (m)	*percentages*						
Fines	Sand	Gravel		Fines	Sand			Gravel		
				$-\frac{1}{16}$	$+\frac{1}{16}-\frac{1}{4}$	$+\frac{1}{4}-1$	+1–4	+4–16	+16–64	+64
3	39	58	3.3–4.3	3	1	13	19	40	24	0
			4.3–5.3	2	2	16	21	40	19	0
			5.3–6.9	4	3	18	22	35	18	0
			Mean	3	2	16	21	38	20	0

COMPOSITION

Depth below surface (m)	Percentage by weight in gravel (+4–64 mm) fraction					
	Limestone including chalk	Flint	Ironstone	Sandstone	Quartzite	Others
3.3–6.9	72	18	10	0	0	0

TF 00 NW 57 **0113 0607** **Welland valley, near Tinwell, Stamford** **Sub-block A_1**

Surface level (+24.4 m) +80 ft
Water struck at (+21.4 m)
152 mm percussion
March 1976

Overburden 3.0 m
Mineral 5.4 m
Bedrock 0.8 m+

LOG

Geological classification	Lithology	Thickness m	Depth m
Made ground	Soil and brick rubble	0.4	0.4
Alluvium	Clay, firm to stiff, fawn mottled pale orange-brown becoming grey towards the base	2.6	3.0
River Terrace Deposits (First Terrace)	Gravel Gravel: fine with some coarse, subangular to subrounded black to dark brown ironstone with subangular to subrounded tabular oolitic and shelly limestone and angular to subangular flint Sand: coarse and medium with trace fine, quartz with limestone ooliths and flint Fines: pale brown to orange-brown	5.4	8.4
Upper Lias	Clay, pebbly in upper 0.2 m. Indurated, dark grey to bluish grey, silty, trace shell fragments some with pyritic veneer, shaly towards base	0.8+	9.2

GRADING

Mean for deposit *percentages*			Depth below surface (m)	*percentages*						
Fines	Sand	Gravel		Fines	Sand			Gravel		
				$-\frac{1}{16}$	$+\frac{1}{16}-\frac{1}{4}$	$+\frac{1}{4}-1$	+1–4	+4–16	+16–64	+64
4	43	53	3.0–4.0	3	2	16	17	39	23	0
			4.0–5.0	3	2	20	20	38	17	0
			5.0–6.0	1	2	17	27	44	9	0
			6.0–7.0	3	3	26	20	41	7	0
			7.0–8.4	6	4	16	25	38	11	0
			Mean	4	2	19	22	40	13	0

COMPOSITION

Depth below surface (m)	Percentage by weight in gravel (+4–64 mm) fraction					
	Limestone including chalk	Flint	Ironstone	Sandstone	Quartzite	Others
3.0–8.4	25	12	60	1	0	2

TF 00 NW 58 0161 0564 Near Dottrell Hill Plantation, Stamford **Sub-block F_3**

Surface level (+25.6 m) +84 ft
Water struck at (+22.7 m and +20.7 m)
152 mm percussion
March 1976

Overburden 0.4 m
Mineral 2.4 m
Waste 2.1 m
Mineral 1.5 m
Waste 1.1 m
Bedrock 2.4 m+

LOG

Geological classification	Lithology	Thickness m	Depth m
	Soil, pale reddish brown with iron-cemented sand and gravel layer at base	0.4	0.4
River Terrace Deposits (First Terrace)	**a** 'Clayey' sandy gravel Gravel: fine with coarse, ironstone with limestone Sand: medium and fine with trace coarse, subangular to subrounded quartz Fines: orange-brown	2.4	2.8
	Clay, firm to stiff, brown to dark grey, with trace pebbles and shell fragments. Sandy in upper 1.7 m	2.1	4.9
	b Gravel Gravel: fine with some coarse, subrounded to subangular oolitic and shelly limestone with ironstone, flint and quartzite Sand: medium and coarse with trace fine, rounded to subrounded, quartz with ironstone and limestone ooliths Fines: pale greyish brown	1.5	6.4
	Silt, firm to stiff, dark greyish blue, with carbonaceous patches, pebbly towards base	1.1	7.5
Upper Lias	Clay, stiff, dark grey to black, silty, occasional indurated siltstone pellets, trace shell fragments, shaly towards base	2.4+	9.9

GRADING

	Mean for deposit percentages			Depth below surface (m)	percentages						
	Fines	Sand	Gravel		Fines	Sand			Gravel		
					$-\frac{1}{16}$	$+\frac{1}{16}-\frac{1}{4}$	$+\frac{1}{4}-1$	+1–4	+4–16	+16–64	+64
a	19	60	21	0.4–1.8	17	23	26	6	19	9	0
				1.8–2.8	20	33	31	2	9	5	0
				Mean	19	28	28	4	14	7	0
				2.8–4.9 waste	46	23	23	5	2	1	0
				4.9–5.9	4	4	24	19	37	13	0
				5.9–6.4	4	4	22	23	38	9	0
b	4	48	48	Mean	4	4	23	21	37	11	0
a+b	13	55	32	Mean	13	18	26	11	23	9	0

COMPOSITION (a+b)

Depth below surface (m)	Percentage by weight in gravel (+4–64 mm) fraction					
	Limestone including chalk	Flint	Ironstone	Sandstone	Quartzite	Others
0.4–6.4	14	9	70	0	4	3

TF 00 NW 59   0213 0599   Near South View Farm, Stamford   Sub-block A_1

Surface level (+24.0 m) +78.5 ft
Water struck at (+20.5 m)
152 mm percussion
March 1976

Overburden 3.5 m
Mineral 0.4 m
Bedrock 1.5 m+

LOG

Geological classification	Lithology	Thickness m	Depth m
	Soil, dark brown	0.2	0.2
Alluvium	Clay, firm to stiff, brown to grey mottle, below 3.0 m becomes black with shell fragments and pebbles	3.3	3.5
River Terrace Deposits (First Terrace)	Gravel Gravel: coarse with fine, mainly subrounded oolitic and shelly limestone with ironstone and subangular to angular flint with occasional subrounded quartzite Sand: medium with coarse and trace fine Fines: dark greyish brown	0.4	3.9
Upper Lias	Clay, stiff, dark grey, pebbly in upper 0.3 m, trace shell fragments	1.5+	5.4

GRADING

Mean for deposit *percentages*			Depth below surface (m)	*percentages*						
Fines	Sand	Gravel		Fines	Sand			Gravel		
				$-\frac{1}{16}$	$+\frac{1}{16}-\frac{1}{4}$	$+\frac{1}{4}-1$	+1–4	+4–16	+16–64	+64
6	30	64	3.5–3.9	6	4	13	13	31	33	0

TF 00 NW 60 0276 0691 Stamford Public Green, Stamford **Sub-block A_1**

Surface level (+22.8 m)+75 ft
Water struck at (+20.5 m)
152 mm percussion
March 1976

Overburden 2.8 m
Mineral 0.8 m
Bedrock 1.0 m+

LOG

Geological classification	Lithology	Thickness m	Depth m
	Soil	0.3	0.3
Alluvium	Silt, soft, pale brown to grey, glutinous, with traces of sand and pebbles, occasional black carbonaceous streaks	2.3	2.6
River Terrace Deposits (First Terrace)	Clay, soft to firm, bluish grey, silty	0.2	2.8
	Gravel Gravel: fine with coarse angular flint with rounded to subrounded shelly and oolitic limestone with subrounded ironstone and quartzite Sand: coarse and medium with trace fine, limestone ooliths with ironstone Fines: dark grey	0.8	3.6
?Lower Estuarine Series	Clay, stiff, dark grey, silty, occasional shell fragments	1.0+	4.6

GRADING

Mean for deposit *percentages*			Depth below surface (m)	*percentages*						
Fines	Sand	Gravel		Fines	Sand			Gravel		
				$-\frac{1}{16}$	$+\frac{1}{16}-\frac{1}{4}$	$+\frac{1}{4}-1$	+1–4	+4–16	+16–64	+64
2	27	71	2.8–3.6	2	1	12	14	40	31	0

COMPOSITION

Depth below surface (m)	Percentage by weight in gravel (+4–64 mm) fraction					
	Limestone including chalk	Flint	Ironstone	Sandstone	Quartzite	Others
2.8–3.6	27	38	20	0	15	0

TF 00 NW 61 0350 0991 South of Ryhall, Stamford **Sub-block F_3**

Surface level (+25.9 m) +85 ft
Water struck at (+20.9 m)
152 mm percussion
March 1976

Overburden 2.7 m
Mineral 2.2 m
Waste 1.2 m
Bedrock 0.5 m+

LOG

Geological classification	Lithology	Thickness m	Depth m
Made ground	Soil overlying clay and rubble	1.1	1.1
River Terrace Deposits (First Terrace)	Clay, mottled khaki-green with black and orange-brown, many limestone pebbles	1.6	2.7
	'Clayey' sandy gravel Gravel: fine with trace coarse, predominantly subangular to subrounded oolitic limestone with angular to subangular flint and subrounded to rounded ironstone Sand: coarse and medium with some fine, mostly limestone ooliths with some flint, ironstone and quartz Fines: orange to pale greyish brown	2.2	4.9
	Clay, pale greyish brown, ?micaceous, with traces of sand and gravel	1.2	6.1
Lower Lincolnshire Limestone	Limestone, weathered, crystalline, pale yellow to cream	0.5+	6.6

GRADING

Mean for deposit percentages			Depth below surface (m)	percentages						
Fines	Sand	Gravel		Fines	Sand			Gravel		
				$-\frac{1}{16}$	$+\frac{1}{16}-\frac{1}{4}$	$+\frac{1}{4}-1$	+1–4	+4–16	+16–64	+64
19	50	31	2.7–3.7	25	13	22	20	17	3	0
			3.7–4.9	13	7	18	21	34	6	0
			Mean	19	10	19	21	26	5	0

COMPOSITION

Depth below surface (m)	Percentage by weight in gravel (+4–64 mm) fraction					
	Limestone including chalk	Flint	Ironstone	Sandstone	Quartzite	Others
2.7–4.9	80	10	8	0	2	0

TF 00 NW 62 0456 0767 South of Newstead Mill, Stamford **Block F**

Surface level (+20.1 m)+66 ft
Water not struck
152 mm percussion
March 1976

Waste 0.6 m
Bedrock 3.1 m+

LOG

Geological classification	Lithology	Thickness m	Depth m
River Terrace Deposits	Soil	0.2	0.2
(First Terrace)	Clay, dark brown mottled orange, trace limestone pebbles	0.4	0.6
Lower Lincolnshire Limestone	Limestone, pale yellowish buff, decalcified upper 0.3 m of sandy ooliths; mostly oolitic with some finely crystalline partings	3.1+	3.7

TF 00 NW 63 0493 0773 South-east of Newstead Mill, Stamford **Sub-block F_3**

Surface level (+20.7 m)+68 ft
Water not struck
152 mm percussion
March 1976

Overburden 0.3 m
Mineral 2.9 m
Bedrock 1.3 m+

LOG

Geological classification	Lithology	Thickness m	Depth m
	Soil, dark greyish brown	0.3	0.3
River Terrace Deposits (First Terrace)	'Clayey' sandy gravel Gravel: fine with some coarse, subrounded to angular limestone with some ironstone, flint and quartzite Sand: medium and coarse with some fine, mainly limestone ooliths with trace flint Fines: pale brown to greyish brown	2.9	3.2
Lower Lincolnshire Limestone	Limestone, pale yellow to buff, oolitic; upper 0.8 m decalcified	1.3+	4.5

GRADING

Mean for deposit *percentages*			Depth below surface (m)	*percentages*						
				Fines	Sand			Gravel		
Fines	Sand	Gravel		$-\frac{1}{16}$	$+\frac{1}{16}-\frac{1}{4}$	$+\frac{1}{4}-1$	+1–4	+4–16	+16–64	+64
15	48	37	0.3–1.3	16	11	21	16	26	10	0
			1.3–2.3	16	11	22	13	33	5	0
			2.3–3.2	13	10	20	20	28	9	0
			Mean	15	11	21	16	29	8	0

COMPOSITION

Depth below surface (m)	Percentage by weight in gravel (+4–64 mm) fraction					
	Limestone including chalk	Flint	Ironstone	Sandstone	Quartzite	Others
0.3–3.2	78	3	11	0	2	6

TF 00 NW 64 **0438 0711** **Welland valley, east of Stamford** **Sub-block F_3**

Surface level (+21.0 m) +69 ft
Water struck at (+18.5 m and +15.4 m)
152 mm percussion
March 1976

Overburden 0.7 m
Mineral 3.7 m
Waste 1.2 m
Mineral 1.4 m
Bedrock 1.0 m+

LOG

Geological classification	Lithology	Thickness m	Depth m
	Soil, dark greyish brown	0.4	0.4
River Terrace Deposits (First Terrace)	Clay, orange-brown mottled dark brown, with pebbles	0.3	0.7
	a 'Clayey' sandy gravel Gravel: fine with trace coarse, subrounded limestone with white and brown angular to subangular flint and subrounded ironstone Sand: medium and fine with trace coarse, quartz, limestone ooliths and ironstone Fines: orange-brown	3.7	4.4
	Clay, soft to firm, orange-brown, silty with sandy pockets	1.2	5.6
	b Gravel Gravel: fine with some coarse, oolitic and crystalline limestone with ironstone and angular to subangular brown flints Sand: coarse and medium with trace fine, mainly limestone ooliths with ironstone and flint Fines: orange-brown	1.4	7.0
?Lower Lincolnshire Limestone	Clay, well indurated, silty, greyish fawn; upper 0.4 m with pebbles	1.0+	8.0

GRADING

	Mean for deposit *percentages*			Depth below surface (m)	*percentages*						
	Fines	Sand	Gravel		Fines	Sand			Gravel		
					$-\frac{1}{16}$	$+\frac{1}{16}-\frac{1}{4}$	$+\frac{1}{4}-1$	+1–4	+4–16	+16–64	+64
a	11	63	26	0.7–1.7	15	21	35	6	19	4	0
				1.7–2.7	7	20	32	7	29	5	0
				2.7–3.7	9	20	30	10	25	6	0
				3.7–4.4	13	33	34	5	11	4	0
				Mean	11	23	33	7	21	5	0
				4.4–5.6 waste							
b	5	44	51	5.6–7.0	5	5	19	20	40	11	0
a+b	9	58	33	Mean	9	18	29	11	27	6	0

COMPOSITION (a+b)

Depth below surface (m)	Percentage by weight in gravel (+4–64 mm) fraction					
	Limestone including chalk	Flint	Ironstone	Sandstone	Quartzite	Others
0.7–7.0	63	16	21	0	0	0

TF 00 NW 66 0029 0641 East of Tinwell, Stamford **Block G**

Surface level (c + 32 m) c + 106 ft
Water not struck
76 mm Minuteman
August 1977

Waste 0.5 m
Bedrock 0.2 m +

LOG

Geological classification	Lithology	Thickness m	Depth m
	Soil, medium brown, passes into sandy clay with limestone pebbles	0.4	0.4
River Terrace Deposits (Second Terrace)	Clay, firm to stiff, pale grey slightly mottled brown	0.1	0.5
Upper Lias	Clay, stiff becoming friable with depth, pale brown to pale grey	0.2 +	0.7

TF 00 NW 67 0039 0621 West of Tinwell, Stamford **Block F**

Surface level unknown
Water not struck
76 mm Minuteman
August 1977

Waste 0.4 m +

LOG

Geological classification	Lithology	Thickness m	Depth m
	Soil, dark brown to grey, clayey with limestone gravel	0.2	0.2
River Terrace Deposits (?First Terrace)	Clay, dark brown with large fragments of oolitic limestone	0.2 +	0.4
	(No drilling progress beyond 0.4 m)		

TF 00 NW 68 0074 0585 Near Tinwell railway crossing, Tinwell **Block F**

Surface level (c + 25 m) c + 83 ft
Water struck at (c + 22 m)
76 mm Minuteman
August 1977

Waste 5.0 m
Bedrock 1.5 m +

LOG

Geological classification	Lithology	Thickness m	Depth m
	Soil, dark greyish brown	0.1	0.1
River Terrace Deposits (First Terrace)	Clay, soft, friable, pale brown to pale greyish brown variegated with orange-brown below 1.0 m	2.9	3.0
	Clay, dark to pale greyish brown, slightly variegated with orange, with fine to coarse limestone and ironstone gravel in upper 1.0 m, becoming sandy below c 4.0 m	2.0	5.0
Upper Lias	Clay, firm to stiff, dark grey to bluish grey micaceous, calcareous	1.5 +	6.5

TF 00 NW 69 0027 0637 West of Tinswell, Stamford **Block G**

Surface level (c + 32 m) c + 106 ft
Water struck at (c + 27 m)
76 mm Minuteman
August 1977

Waste 6.0 m
Bedrock 1.0 m +

LOG

Geological classification	Lithology	Thickness m	Depth m
	Soil, dark greyish brown, trace limestone pebbles	0.5	0.5
River Terrace Deposits (Second Terrace)	Clay, pale orange-brown, slightly calcareous, trace oolitic and shelly limestone gravel with fine quartz and ironstone sand	5.5	6.0
Upper Lias	Clay, firm to stiff, pale brownish grey to dark greyish blue, ?micaceous, calcareous	1.0 +	7.0

TF 00 NE 43 0514 0717 Welland valley, east of Stamford **Sub-block A_1**

Surface level (+ 18.0 m) + 59 ft
Water level not recorded
152 mm percussion
March 1977

Overburden 0.4 m
Mineral 1.0 m
Bedrock 4.9 m +

LOG

Geological classification	Lithology	Thickness m	Depth m
Alluvium	Soil, brown, clayey with limestone gravel	0.4	0.4
River Terrace Deposits (First Terrace)	Gravel Gravel: mainly fine with coarse and some cobble, subrounded to rounded crystalline oolitic and shelly limestone with sandstone and ironstone Sand: coarse and medium with trace fine, limestone ooliths and quartz Fines: generally pale yellowish brown	1.0	1.4
Lower Lincolnshire Limestone	Limestone, oolitic, upper surface of broken and crushed rock	4.9 +	6.3

GRADING

Mean for deposit *percentages*			Depth below surface (m)	*percentages*						
Fines	Sand	Gravel		Fines	Sand			Gravel		
				$-\frac{1}{16}$	$+\frac{1}{16}-\frac{1}{4}$	$+\frac{1}{4}-1$	+1–4	+4–16	+16–64	+64
8	38	54	0.4–1.4	8	5	16	17	35	12	7

TF 00 NE 44 0506 0728 Uffington Park, Stamford **Sub-block A_1**

Surface level (+18.9 m) +62 ft
Water not struck
152 mm percussion
March 1976

Overburden 2.6 m
Mineral 1.9 m
Bedrock 2.2 m+

LOG

Geological classification	Lithology	Thickness m	Depth m
Alluvium	Soil, dark brown	0.2	0.2
	Clay, firm to stiff, pale brown becomes grey towards base, silty	2.4	2.6
River Terrace Deposits (First Terrace)	"Clayey' sandy gravel Gravel: mainly fine with some coarse and trace cobble, mostly oolitic and crystalline limestone with ironstone and some flint and quartzite Sand: medium and coarse with some fine, quartz with ironstone and limestone oolitic Fines: dark grey to black	1.9	4.5
Lower Lincolnshire Limestone	Limestone, oolitic, pale yellowish cream to cream; decalcified upper sandy margin passing into well indurated bedded rock	2.2+	6.7

GRADING

Mean for deposit percentages			Depth below surface (m)	percentages						
Fines	Sand	Gravel		Fines	Sand			Gravel		
				$-\frac{1}{16}$	$+\frac{1}{16}-\frac{1}{4}$	$+\frac{1}{4}-1$	+1–4	+4–16	+16–64	+64
12	45	43	2.6–3.6	12	8	22	15	31	10	2
			3.6–4.5	12	8	21	16	33	10	0
			Mean	12	8	21	16	32	10	1

COMPOSITION

Depth below surface (m)	Percentage by weight in gravel (+4–64 mm) fraction					
	Limestone including chalk	Flint	Ironstone	Sandstone	Quartzite	Others
2.6–4.5	50	18	24	1	7	0

TF 00 NE 45 0542 0675 Welland valley, east Stamford **Sub-block F_3**

Surface level (+19.2 m) +63 ft
Water not struck
152 mm percussion
March 1976

Overburden 0.6 m
Mineral 5.4 m
Waste 0.7 m
Bedrock 1.3 m+

LOG

Geological classification	Lithology	Thickness m	Depth m
Made ground	Brick and limestone rubble with soil	0.4	0.4
River Terrace Deposits (First Terrace)	Clay, brown, silty, trace pebble gravel	0.2	0.6
	Sandy gravel Gravel: mainly fine with coarse and trace cobble, mostly limestone with ironstone and brown angular to subangular flint and some rounded to subrounded sandstone Sand: mainly medium with coarse and some fine, mostly quartz and limestone ooliths Fines: pale yellowish brown	5.4	6.0
	Clay, mottled pale brown with orange with traces of ironstone, sandstone and limestone gravel	0.7	6.7
Lower Lincolnshire Limestone	Limestone, oolitic, well indurated, pale yellow to brown; decalcified in parts giving oolitic sand	1.3+	8.0

GRADING

Mean for deposit *percentages* Fines	Sand	Gravel	Depth below surface (m)	*percentages* Fines $-\frac{1}{16}$	Sand $+\frac{1}{16}-\frac{1}{4}$	$+\frac{1}{4}-1$	$+1-4$	Gravel $+4-16$	$+16-64$	$+64$
9	49	42	0.6–1.6	9	8	29	12	36	5	0
			1.6–2.6	9	8	32	12	30	9	0
			2.6–3.6	10	9	31	13	30	7	0
			3.6–4.6	9	12	26	15	30	8	0
			4.6–6.0	10	7	16	15	20	17	7
			Mean	9	9	26	14	30	10	2

COMPOSITION

Depth below surface (m)	Percentage by weight in gravel (+4–64 mm) fraction: Limestone including chalk	Flint	Ironstone	Sandstone	Quartzite	Others
0.6–6.0	54	14	25	5	2	0

TF 00 NE 46 0592 0635 The Butlands, east Stamford **Sub-block F_3**

Surface level (+20.1 m) +66 ft
Water not struck
152 mm percussion
March 1976

Overburden 1.3 m
Mineral 1.5 m
Waste 0.4 m
Mineral 0.8 m
Bedrock 0.7 m+

LOG

Geological classification	Lithology	Thickness m	Depth m
	Soil, brown to dark grey	0.4	0.4
River Terrace Deposits (First Terrace)	Clay, firm, brown, silty and sandy in part	0.9	1.3
	a 'Clayey' sandy gravel Gravel: fine with coarse, trace cobble, angular to subrounded oolitic and crystalline limestone and ironstone, with flint and some quartzite Sand: mainly medium with fine and some coarse, quartz with limestone ooliths Fines: orange-brown, below 2.3 m becomes yellowish grey	1.5	2.8
	Clay, stiff, silty, micaceous, iron-oxide stained, with trace gravel	0.4	3.2
	b 'Clayey' sandy gravel—as above	0.8	4.0
Lower Lincolnshire Limestone	Limestone, pale yellowish cream to white; upper part mainly decalcified limestone–clay mixture, passes into broken rock	0.7+	4.7

GRADING

	Mean for deposit *percentages*			Depth below surface (m)	*percentages*						
	Fines	Sand	Gravel		Fines	Sand			Gravel		
					$-\frac{1}{16}$	$+\frac{1}{16}-\frac{1}{4}$	$+\frac{1}{4}-1$	+1–4	+4–16	+16–64	+64
a	11	66	23	1.3–2.3	11	21	42	7	17	3	0
				2.3–2.8	13	13	33	12	24	5	0
				Mean	11	18	39	9	19	4	0
					2.8–3.2 waste						
b	15	56	29	3.2–4.0	15	14	32	10	14	9	6
a+b	13	62	25	Mean	13	17	36	9	17	6	2

COMPOSITION

Depth below surface (m)	Percentage by weight in gravel (+4–64 mm) fraction					
	Limestone including chalk	Flint	Ironstone	Sandstone	Quartzite	Others
1.3–4.0	40	17	40	0	3	0

TF 00 NE 47 **0648 0727** **Uffington Park Lodge, Stamford** **Block O**

Surface level (+32.3 m) + 106 ft
Water struck at (+29.7 m)
152 mm percussion
March 1976

Overburden 0.1 m
Mineral 4.1 m
Bedrock 3.6 m+

LOG

Geological classification	Lithology	Thickness m	Depth m
	Soil, dark brown	0.1	0.1
Glacial Sand and Gravel	'Clayey' sandy gravel—'very clayey' in upper 1.0 m Gravel: fine and coarse with trace cobble, rounded to subrounded oolitic limestone with angular to subangular flint, and some ironstone and quartzite Sand: medium and coarse with some fine, quartz with limestone ooliths and ironstone Fines: dark to pale greyish brown	4.1	4.2
Kellaways Sand	Silt, firm to stiff becoming soft below 5.0 m, pale fawnish grey to grey mottled with orange-brown	2.6	6.8
Kellaways Clay	Clay, stiff, dark grey with silty patches	1.0+	7.8

GRADING

Mean for deposit *percentages*			Depth below surface (m)	*percentages*						
Fines	Sand	Gravel		Fines	Sand			Gravel		
				$-\frac{1}{16}$	$+\frac{1}{16}-\frac{1}{4}$	$+\frac{1}{4}-1$	+1–4	+4–16	+16–64	+64
14	47	39	0.1–1.1	33	14	22	7	14	10	0
			1.1–2.1	12	12	27	9	28	12	0
			2.1–3.0	7	5	18	16	33	19	2
			3.0–4.2	7	9	19	29	20	16	0
			Mean	14	10	22	15	24	14	1

COMPOSITION

Depth below surface (m)	Percentage by weight in gravel (+4–64 mm) fraction					
	Limestone including chalk	Flint	Ironstone	Sandstone	Quartzite	Others
0.1–4.2	68	20	10	0	2	0

TF 00 NE 48 0756 0925 Near Casewick Hall, Stamford **Block O**

Surface level (+25.9 m) +85 ft
Water not struck
152 mm percussion
April 1976

Waste 4.5 m
Bedrock 0.3 m+

LOG

Geological classification	Lithology	Thickness m	Depth m
	Soil, grey brown, sandy	0.8	0.8
?Glacial Sand and Gravel	Clay, firm to stiff, variegated pale orange-brown to grey, silty, calcareous with decalcified limestone pebbles; thin silty layer at 2.4 m	3.7	4.5
Kellaways Sand	Siltstone, pale brown to buff, calcareous	0.3+	4.8

TF 00 NE 49 0761 0822 South of Casewick Park, Stamford **Block O**

Surface level (+22.3 m) +73 ft
Water not struck
152 mm percussion
March 1976

Waste 1.7 m
Bedrock 1.7 m+

LOG

Geological classification	Lithology	Thickness m	Depth m
	Soil, greyish brown	0.3	0.3
Glacial Sand and Gravel	Clay, stiff pale yellow to grey, very pebbly with limestone, sandstone and flint gravel and trace shell fragments	1.4	1.7
Oxford Clay	Clay, stiff variegated dark bluish grey, orange and dark brown; indurated siltstone at base	1.7+	3.4

TF 00 NE 50 0732 0743 Copthill Farm, Uffington **Block O**

Surface level (+31.4 m)+103 ft
Water struck at (+29.4 m)
152 mm percussion
March 1976

Overburden 0.1 m
Mineral 2.7 m
Bedrock 0.8 m+

LOG

Geological classification	Lithology	Thickness m	Depth m
	Soil, sandy	0.1	0.1
Glacial Sand and Gravel	'Clayey' sandy gravel Gravel: fine and coarse with trace cobble, rounded to subrounded oolitic limestone with angular flint and rounded ironstone Sand: coarse and medium with fine, quartz with limestone ooliths Fines: pale brown	2.7	2.8
Oxford Clay	Clay, stiff, pale greyish brown becoming dark grey below 3.1 m, shaly, with numerous bivalve impressions	0.8+	3.6

GRADING

Mean for deposit *percentages*			Depth below surface (m)	*percentages*						
Fines	Sand	Gravel		Fines	Sand			Gravel		
				$-\frac{1}{16}$	$+\frac{1}{16}-\frac{1}{4}$	$+\frac{1}{4}-1$	+1–4	+4–16	+16–64	+64
16	43	41	0.1–1.1	27	12	14	26	16	6	0
			1.1–2.1	11	6	17	16	32	16	2
			2.1–2.8	7	5	12	21	36	16	3
			Mean	16	8	16	19	27	12	2

COMPOSITION

Depth below surface (m)	Percentage by weight in gravel (+4–64 mm) fraction					
	Limestone including chalk	Flint	Ironstone	Sandstone	Quartzite	Others
0.1–2.8	71	12	14	0	3	0

TF 00 NE 51 **0702 0679** **Near Pilsgate Grange, Barnack** **Sub-block A_1**

Surface level (+16.5 m) +54 ft
Water struck at (+14.3 m)
152 mm percussion
March 1976

Overburden 2.2 m
Mineral 4.1 m
Bedrock 0.3 m+

LOG

Geological classification	Lithology	Thickness m	Depth m
	Soil, dark greyish brown	0.2	0.2
Alluvium	Clay, soft to firm becoming stiff below 0.1 m, brown to grey, silty	2.0	2.2
River Terrace Deposits (First Terrace)	Gravel Gravel: fine and coarse with trace cobble, subrounded to rounded, shelly and oolitic limestone with ironstone and with some angular to subangular flint Sand: coarse and medium with fine, limestone ooliths with ironstone, flint and quartz Fines: pale brown to greyish brown	4.1	6.3
Upper Estuarine 'Series'	Limestone, shelly, well indurated, pale grey to dark grey	0.3+	6.6

GRADING

Mean for deposit percentages			Depth below surface (m)	percentages						
Fines	Sand	Gravel		Fines	Sand			Gravel		
				$-\frac{1}{16}$	$+\frac{1}{16}-\frac{1}{4}$	$+\frac{1}{4}-1$	+1–4	+4–16	+16–64	+64
4	39	57	2.2–3.2	5	3	17	19	37	14	5
			3.2–4.2	3	2	17	24	31	23	0
			4.2–6.3	3	3	15	20	35	24	0
			Mean	4	2	16	21	34	22	1

COMPOSITION

Depth below surface (m)	Percentage by weight in gravel (+4–64 mm) fraction					
	Limestone including chalk	Flint	Ironstone	Sandstone	Quartzite	Others
2.2–6.3	45	18	28	0	8	1

TF 00 NE 52 **0765 0618** **North of Barnack** **Sub-block F_1**

Surface level (+16.5 m) +54 ft
Water struck at (+12.0 m)
152 mm percussion
March 1976

Waste 4.3 m
Bedrock 0.3 m+

LOG

Geological classification	Lithology	Thickness m	Depth m
	Soil, brown	0.4	0.4
River Terrace Deposits (First Terrace)	Clay, soft to firm, brown mottled with orange-brown and pale grey, sandy with some pebbles, shell fragments and carbonaceous patches	3.9	4.3
Kellaways Clay	Siltstone, well indurated, dark grey, calcareous	0.3	4.6

TF 00 NE 53 0837 0975 Casewick Park, Stamford **Sub-block F_2**

Surface level (+14.0 m) +46 ft
Water struck at (+11.9 m)
152 mm percussion
April 1976

Overburden 0.5 m
Mineral 0.8 m
Bedrock 1.1 m+

LOG

Geological classification	Lithology	Thickness m	Depth m
	Soil, greyish brown	0.5	0.5
River Terrace Deposits (First Terrace)	Gravel Gravel: mainly fine with some coarse and trace cobble, subrounded to rounded, oolitic and shelly limestone with angular to subangular flints, rounded to subrounded ironstone and quartzite Sand: medium and coarse with trace fine, quartz, ironstone, flint with limestone ooliths Fines: pale greyish-brown	4.1	4.6
Oxford Clay	Clay, weathered in upper 0.3 m, firm to stiff, dark to pale grey, micaceous in part; indurated at base? siltstone	1.1+	5.7

GRADING

Mean for deposit percentages			Depth below surface (m)	percentages						
Fines	Sand	Gravel		Fines	Sand			Gravel		
				$-\frac{1}{16}$	$+\frac{1}{16}-\frac{1}{4}$	$+\frac{1}{4}-1$	+1–4	+4–16	+16–64	+64
4	39	57	0.5–1.5	7	7	27	16	39	4	0
			1.5–2.5	2	2	10	17	45	17	7
			2.5–3.5	2	3	15	18	47	15	0
			3.5–4.6	3	4	19	17	39	17	1
			Mean	4	4	18	17	42	13	2

COMPOSITION

Depth below surface (m)	Percentage by weight in gravel (+4–64 mm) fraction					
	Limestone including chalk	Flint	Ironstone	Sandstone	Quartzite	Others
0.5–4.6	53	21	19	0	7	0

Surface level (+14.3 m) +47 ft
Water struck at (+13.2 m)
152 mm percussion
April 1976

Overburden 0.4 m
Mineral 2.7 m
Bedrock 0.7+

LOG

Geological classification	Lithology	Thickness m	Depth m
	Soil, brown	0.2	0.2
River Terrace Deposits	Clay, firm, pale yellowish brown, silty sand towards base	0.2	0.4
(First Terrace)	Gravel—'clayey' in upper 0.2 m Gravel: mainly fine with some coarse and trace cobble, subrounded to rounded, tabular, oolitic limestone with flint and ironstone Sand: medium and coarse with some fine, quartz, limestone ooliths and ironstone Fines: pale brown	2.7	3.1
Kellaways Clay	Limestone, shelly, indurated, pale yellowish brown becoming dark grey, with a siltstone band between 3.6–3.7 m	0.7+	3.8

GRADING

Mean for deposit *percentages*			Depth below surface (m)	*percentages*						
Fines	Sand	Gravel		Fines	Sand			Gravel		
				$-\frac{1}{16}$	$+\frac{1}{16}-\frac{1}{4}$	$+\frac{1}{4}-1$	+1–4	+4–16	+16–64	+64
5	46	49	0.4–1.4	8	12	24	17	33	6	0
			1.4–2.4	3	9	17	16	41	14	5
			2.4–3.1	4	5	22	20	38	10	2
			Mean	5	8	21	17	37	10	2

COMPOSITION

Depth below surface (m)	Percentages by weight in gravel (+4–64 mm) fraction					
	Limestone including chalk	Flint	Ironstone	Sandstone	Quartzite	Others
0.4–3.1	58	24	16	1	1	0

TF 00 NE 55 0828 0775 West of Tallington, Stamford **Sub-block F_2**

Surface level (+16.5 m)+54 ft
Water struck at (+14.1 m)
152 mm percussion
March 1976

Overburden 0.4 m
Mineral 5.6 m
Bedrock 0.5 m+

LOG

Geological classification	Lithology	Thickness m	Depth m
	Soil, brown, sandy	0.4	0.4
River Terrace Deposits (First Terrace)	'Clayey' sandy gravel Gravel: mainly fine with some coarse, rounded oolitic limestone with ironstone, angular flint and some quartzite Sand: coarse and medium with trace fine, quartz, limestone ooliths, ironstone and flint Fines: pale brown	5.6	6.0
Kellaways Clay	Clay, stiff, dark grey with shell impressions	0.5+	6.5

GRADING

Mean for deposit *percentages*			Depth below surface (m)	*percentages*						
Fines	Sand	Gravel		Fines	Sand			Gravel		
				$-\frac{1}{16}$	$+\frac{1}{16}-\frac{1}{4}$	$+\frac{1}{4}-1$	+1–4	+4–16	+16–64	+64
10	45	45	0.4–0.8	38	24	30	4	3	1	0
			0.8–1.8	10	8	28	14	35	5	0
			1.8–2.8	10	8	24	13	27	18	0
			2.8–3.8	1	3	17	20	38	21	0
			3.8–4.8	4	2	14	23	40	17	0
			4.8–6.0	3	5	20	19	39	14	0
			Mean	10	8	21	16	32	13	0

COMPOSITION

Depth below surface (m)	Percentages by weight in gravel (+4–64 mm) fraction					
	Limestone including chalk	Flint	Ironstone	Sandstone	Quartzite	Others
0.4–6.0	63	12	19	0	5	1

Surface level (+15.5 m) +51 ft
Water struck at (+13.5 m)
152 mm percussion
March 1976

Overburden 2.0 m
Mineral 4.5 m
Bedrock 0.6 m+

LOG

Geological classification	Lithology	Thickness m	Depth m
	Soil, dark greyish brown	0.3	0.3
Alluvium	Clay, firm, pale brown to pale grey, silty in part	1.7	2.0
River Terrace Deposits (First Terrace)	Gravel Gravel: mainly fine with coarse, pale yellow, rounded to subrounded, tabular, oolitic and shelly limestone and black to red, subrounded ironstone with angular to subangular flint and some quartzite Sand: coarse and medium with trace fine, quartz, limestone ooliths, ironstone and some flint Fines: pale greyish brown	4.5	6.5
?Cornbrash	Clay, stiff to very stiff, dark bluish grey with pyrite and shell fragments	0.6+	7.1

GRADING

Mean for deposit percentages: Fines	Sand	Gravel	Depth below surface (m)	percentages: Fines $-\frac{1}{16}$	Sand $+\frac{1}{16}-\frac{1}{4}$	Sand $+\frac{1}{4}-1$	Sand $+1-4$	Gravel $+4-16$	Gravel $+16-64$	Gravel $+64$
4	36	60	2.0–3.0	4	1	15	15	36	29	0
			3.0–4.0	4	3	17	22	40	14	0
			4.0–5.0	5	3	15	20	41	16	0
			5.0–6.5	4	2	10	21	43	20	0
			Mean	4	2	14	20	40	20	0

COMPOSITION

Depth below surface (m)	Percentage by weight in gravel (+4–64 mm) fraction: Limestone including chalk	Flint	Ironstone	Sandstone	Quartzite	Others
2.0–6.5	36	19	31	2	12	0

TF 00 NE 57 **0842 0653** **North of Barnack, Stamford** **Sub-block F_1**

Surface level (+16.2 m) +53 ft
Water struck at (+13.7 m)
152 mm percussion
March 1976

Overburden 0.9 m
Mineral 5.1 m
Bedrock 2.0 m+

LOG

Geological classification	Lithology	Thickness m	Depth m
	Soil, brown	0.4	0.4
River Terrace Deposits	Clay, firm to stiff, brown to dark brown, sandy	0.5	0.9
(First Terrace)	Gravel	5.1	6.0
	Gravel: mainly fine with coarse, mostly subrounded to subangular oolitic limestone, with subrounded ironstone and angular to subangular flint		
	Sand: medium and coarse with trace fine, quartz with limestone ooliths and ironstone		
	Fines: dark orange-brown		
Kellaways Clay	Clay, stiff, dark bluish grey, trace selenite and fossils fragments	2.0+	8.0

GRADING

Mean for deposit *percentages*			Depth below surface (m)	*percentages*						
Fines	Sand	Gravel		Fines	Sand			Gravel		
				$-\frac{1}{16}$	$+\frac{1}{16}-\frac{1}{4}$	$+\frac{1}{4}-1$	+1–4	+4–16	+16–64	+64
5	44	51	0.9–1.9	7	11	29	11	38	4	0
			1.9–2.9	9	9	26	21	31	4	0
			2.9–3.9	3	4	23	19	43	8	0
			3.9–4.9	3	3	22	19	39	14	0
			4.9–6.0	4	1	8	17	39	31	0
			Mean	5	6	21	17	38	13	0

COMPOSITION

Depth below surface (m)	Percentage by weight in gravel (+4–64 mm) fraction					
	Limestone including chalk	Flint	Ironstone	Sandstone	Quartzite	Others
0.9–6.0	51	18	23	1	5	2

TF 00 NE 58 0837 0551 East of Barnack, Stamford **Sub-block F_1**

Surface level (+20.1 m) +66 ft
Water struck at (+17.8 m)
152 mm percussion
March 1976

Overburden 1.3 m
Mineral 3.6 m
Bedrock 0.9 m+

LOG

Geological classification	Lithology	Thickness m	Depth m
?Made ground	Soil, yellowish brown, clay with stone fragments	1.0	1.0
River Terrace Deposits	Clay, orange-brown to brown with flint and sandstone pebbles	0.3	1.3
(First Terrace)	Gravel	3.6	4.9
	Gravel: mainly fine with some coarse, rounded to subrounded, tabular, oolitic limestone with angular to subangular ironstained flint and some ironstone and quartzite		
	Sand: medium and coarse with trace fine, quartz, ironstone and sandstone		
	Fines: orange-brown		
Oxford	Clay, stiff, dark to pale grey, with shell fragments	0.9+	5.8

GRADING

Mean for deposit percentages			Depth below surface (m)	*percentages*						
Fines	Sand	Gravel		Fines	Sand			Gravel		
				$-\frac{1}{16}$	$+\frac{1}{16}-\frac{1}{4}$	$+\frac{1}{4}-1$	+1–4	+4–16	+16–64	+64
5	41	54	1.3–2.3	6	5	20	21	41	7	0
			2.3–3.3	5	5	23	16	45	6	0
			3.3–4.9	5	4	18	14	40	19	0
			Mean	5	5	20	16	42	12	0

COMPOSITION

Depth below surface (m)	Percentage by weight in gravel (+4–64 mm) fraction					
	Limestone including chalk	Flint	Ironstone	Sandstone	Quartzite	Others
1.3–4.9	55	21	13	0	9	2

TF 00 NE 59 **0924 0931** **Rabbi Cottage, Tallington** **Sub-block F_2**

Surface level (+13.1 m) +43 ft
Water struck at (+11.6 m)
152 mm percussion
March 1976

Overburden 0.4 m
Mineral 2.2 m
Bedrock 0.7 m+

LOG

Geological classification	Lithology	Thickness m	Depth m
	Soil, brown, sandy	0.4	0.4
River Terrace Deposits (First Terrace)	'Clayey' sandy gravel Gravel: fine with coarse, rounded oolitic limestone and angular flint with ironstone and quartzite Sand: medium with fine and coarse, quartz, limestone ooliths and ironstone Fines: pale brown	2.2	2.6
Oxford Clay	Clay, grey, silty with shell fragments	0.7+	3.3

GRADING

Mean for deposit percentages			Depth below surface (m)	percentages						
Fines	Sand	Gravel		Fines	Sand			Gravel		
				$-\frac{1}{16}$	$+\frac{1}{16}-\frac{1}{4}$	$+\frac{1}{4}-1$	+1–4	+4–16	+16–64	+64
10	57	33	0.4–1.4	11	11	27	15	30	6	0
			1.4–1.8	7	11	46	12	22	2	0
			Mean	10	11	32	14	28	5	0

COMPOSITION

Depth below surface (m)	Percentage by weight in gravel (+4–64 mm) fraction					
	Limestone including chalk	Flint	Ironstone	Sandstone	Quartzite	Others
0.4–1.8	48	37	12	1	2	0

TF 00 NE 60 **0896 0824** **Near Tallington, Stamford** **Sub-block F_2**

Surface level (+14.9 m) +49 ft
Water struck at (+13.7 m)
152 mm percussion
February 1976

Waste 1.9 m
Bedrock 0.3 m+

LOG

Geological classification	Lithology	Thickness m	Depth m
	Soil, brown, sandy	0.3	0.3
River Terrace Deposits (First Terrace)	Clay, pale to dark brown, sandy becomes pebbly towards base	1.6	1.9
Kellaways Clay	Siltstone, indurated, greyish brown to greyish green, with thin clay bands and shell fragments	0.3+	2.2

TF 00 NE 61 0919 0768 Church Meadow, Tallington **Sub-block A_1**

Surface level (+14.3 m) +47 ft
Water struck at (+11.4 m)
152 mm percussion
February 1976

Overburden 2.9 m
Mineral 1.1 m
Bedrock 0.6 m+

LOG

Geological classification	Lithology	Thickness m	Depth m
	Soil, brown, sandy	0.3	0.3
Alluvium	Clay, stiff, pale brown becomes bluish grey below 1.8 m, occasional pebbles and traces of carbonaceous material	2.6	2.9
River Terrace Deposits (First Terrace)	Gravel Gravel: fine with some coarse, subrounded ironstone, angular flint, subrounded oolitic limestone with sandstone and quartzite Sand: coarse and medium with trace fine, flint, ironstone, quartz and limestone ooliths Fines: pale brown	1.1	4.0
Kellaways Clay	Clay, stiff, fissured, dark grey, trace shell impressions	0.6+	4.6

GRADING

Mean for deposit *percentages*			Depth below surface (m)	*percentages*						
Fines	Sand	Gravel		Fines	Sand			Gravel		
				$-\frac{1}{16}$	$+\frac{1}{16}-\frac{1}{4}$	$+\frac{1}{4}-1$	+1–4	+4–16	+16–64	+64
2	37	61	2.9–4.1	2	1	11	25	40	21	0

COMPOSITION

Depth below surface (m)	Percentage by weight in gravel (+4–64 mm) fraction					
	Limestone including chalk	Flint	Ironstone	Sandstone	Quartzite	Others
2.9–4.1	28	30	31	2	9	0

TF 00 NE 62 **0945 0718** **North of Bainton, Stamford** **Sub-block F_1**

Surface level (+14.3 m) +47 ft
Water struck at (+12.1 m)
152 mm percussion
February 1976

Overburden 1.0 m
Mineral 3.4 m
Bedrock 1.2 m+

LOG

Geological classification	Lithology	Thickness m	Depth m
	Soil, brown, sandy	0.3	0.3
River Terrace Deposits (First Terrace)	Clay, stiff, pale brown, sandy, pebbly towards base	0.7	1.0
	Sandy gravel Gravel: fine with some coarse, rounded to subrounded oolitic limestone with subrounded ironstone, angular flint and subrounded quartzite Sand: coarse and medium with trace fine, ironstone, flint, quartz and limestone ooliths Fines: brown to pale brown	3.4	4.4
Kellaways Clay	Clay, stiff, fissured, dark grey, silty, laminated	1.2+	5.6

GRADING

Mean for deposit *percentages*			Depth below surface (m)	*percentages*						
Fines	Sand	Gravel		Fines	Sand			Gravel		
				$-\frac{1}{16}$	$+\frac{1}{16}-\frac{1}{4}$	$+\frac{1}{4}-1$	+1–4	+4–16	+16–64	+64
8	46	46	1.0–1.6	15	7	27	24	26	1	0
			1.6–2.6	12	7	27	14	30	10	0
			2.6–3.6	4	3	14	25	44	10	0
			3.6–4.4	3	2	14	24	34	23	0
			Mean	8	5	20	21	35	11	0

COMPOSITION

Depth below surface (m)	Percentage by weight in gravel (+4–64 mm) fraction					
	Limestone including chalk	Flint	Ironstone	Sandstone	Quartzite	Others
1.0–4.4	42	13	29	1	14	1

TF 00 NE 63   **0955 0645**   **Near Bainton, Stamford**   **Sub-block F_1**

Surface level (+15.5 m) +51 ft
Water struck at (+12.2 m)
152 mm percussion
February 1976

Overburden 0.2 m
Mineral 4.1 m
Waste 1.0 m
Bedrock 0.7 m+

LOG

Geological classification	Lithology	Thickness m	Depth m
	Soil, brown, sandy	0.2	0.2
River Terrace Deposits (First Terrace)	'Clayey' sandy gravel Gravel: fine with trace coarse, rounded to subrounded oolitic limestone with subrounded ironstone, subangular brown flint and some rounded to subrounded quartzite and sandstone Sand: medium with coarse and some fine, quartz with limestone ooliths Fines: brown to pale brown	4.1	4.3
	Clay, brown, sandy, with oolitic limestone and ironstone pebbles	1.0	5.3
Kellaways Clay	Clay, stiff to indurated, dark grey, silty	0.7+	6.0

GRADING

Mean for deposit *percentages*			Depth below surface (m)	*percentages*						
Fines	Sand	Gravel		Fines	Sand			Gravel		
				$-\frac{1}{16}$	$+\frac{1}{16}-\frac{1}{4}$	$+\frac{1}{4}-1$	+1–4	+4–16	+16–64	+64
13	55	32	0.2–1.3	27	23	34	6	8	2	0
			1.3–2.3	8	10	33	12	34	3	0
			2.3–3.3	11	12	29	16	29	3	0
			3.3–4.3	4	6	16	21	46	7	0
			Mean	13	13	28	14	28	4	0

COMPOSITION

Depth below surface (m)	Percentages by weight in gravel (+4–64 mm) fraction					
	Limestone including chalk	Flint	Ironstone	Sandstone	Quartzite	Others
0.2–4.3	43	14	28	4	11	0

TF 00 NE 64   0913 0587   South of Bainton, Stamford   **Sub-block F_1**

Surface level (+17.7 m) +58 ft
Water struck at (+16.1 m)
152 mm percussion
February 1976

Overburden 1.0 m
Mineral 1.6 m
Waste 0.5 m
Bedrock 0.2 m+

LOG

Geological classification	Lithology	Thickness m	Depth m
	Soil, brown, sandy, trace pebbles	0.2	0.2
River Terrace Deposit (First Terrace)	Clay, pale brown, sandy, pebbly	0.8	1.0
	'Very clayey' gravel Gravel: fine with some coarse, mainly subrounded oolitic limestone with angular brown flint and rounded ironstone and quartzite Sand: mainly medium and fine with some coarse, quartz, limestone ooliths, flint and ironstone Fines: pale brown to yellow	1.6	2.6
?Oxford Clay	Clay, greyish brown, sandy with pebbles	0.5	3.1
Oxford Clay	Clay, greyish green, silty, with bivalve shell and belemnite fragments	0.2+	3.3

GRADING

Mean for deposit *percentages*			Depth below surface (m)	*percentages*						
Fines	Sand	Gravel		Fines	Sand			Gravel		
				$-\frac{1}{16}$	$+\frac{1}{16}-\frac{1}{4}$	$+\frac{1}{4}-1$	+1–4	+4–16	+16–64	+64
20	38	42	1.0–2.2	14	12	18	10	37	9	0
			2.2–2.6	37	12	13	8	24	6	0
			Mean	20	12	17	9	34	8	0

COMPOSITION

Depth below surface (m)	Percentage by weight in gravel (+14–64 mm) fraction					
	Limestone including chalk	Flint	Ironstone	Sandstone	Quartzite	Others
1.0–2.6	57	21	9	0	9	4

TF 00 NE 65 0984 0812 Mill Road, Tallington **Sub-block F_2**

Surface level (+13.7 m) +45 ft
Water struck at (+12.3 m)
152 mm percussion
February/March 1976

Overburden 0.9 m
Mineral 1.3 m
Bedrock 1.3 m+

LOG

Geological classification	Lithology	Thickness m	Depth m
	Soil, brown, sandy	0.1	0.1
River Terrace Deposits (First Terrace)	Clay, brown, sandy with some pebbles	0.8	0.9
	Sandy gravel Gravel: fine with some coarse, mainly rounded oolitic limestone with angular flint, subrounded to rounded ironstone and quartzite Sand: medium and coarse with some fine, quartz with limestone ooliths Fines: pale brown	1.3	2.2
Kellaways Clay	Clay, weathered in upper 0.7 m, indurated, grey, silty, with shell impressions below 2.9 m	1.3+	3.5

GRADING

Mean for deposit *percentages* Fines	Sand	Gravel	Depth below surface (m)	*percentages* Fines $-\frac{1}{16}$	Sand $+\frac{1}{16}-\frac{1}{4}$	$+\frac{1}{4}-1$	$+1-4$	Gravel $+4-16$	$+16-64$	$+64$
6	54	40	0.9–1.4	10	14	42	12	18	4	0
			1.4–2.2	4	5	20	20	40	11	0
			Mean	6	9	28	17	32	8	0

COMPOSITION

Depth below surface (m)	Percentage by weight in gravel (+4–64 mm) fraction: Limestone including chalk	Flint	Ironstone	Sandstone	Quartzite	Others
0.9–2.2	56	11	16	0	17	0

TF 00 NE 66 0981 0725 North of Bainton, Stamford

Surface level (+13.7 m) +45 ft
Water struck at (+11.7 m)
152 mm percussion
February 1976

Overburden 1.1 m
Mineral 2.9 m
Bedrock 1.4 m+

LOG

Geological classification	Lithology	Thickness m	Depth m
	Soil, brown, sandy	0.3	0.3
River Terrace Deposits (First Terrace)	Clay, pale brown, silty, sandy, pebbly towards base	0.8	1.1
	Sandy gravel Gravel: fine with some coarse, mainly subrounded oolitic limestone with ironstone and angular to subangular flint with rounded quartzite Sand: medium and coarse with trace fine, quartz and ironstone Fines: pale brown	2.9	4.0
Kellaways Clay	Clay, indurated, dark grey, silty, with ammonite impressions	1.4+	5.4

GRADING

Mean for deposit percentages			Depth below surface (m)	percentages						
Fines	Sand	Gravel		Fines	Sand			Gravel		
				$-\frac{1}{16}$	$+\frac{1}{16}-\frac{1}{4}$	$+\frac{1}{4}-1$	+1–4	+4–16	+16–64	+64
7	49	44	1.1–1.3	13	8	20	16	31	12	0
			1.3–2.3	11	5	25	20	36	3	0
			2.3–3.3	3	4	23	19	40	11	0
			3.3–4.0	5	7	23	22	35	8	0
			Mean	7	6	23	20	36	8	0

COMPOSITION

Depth below surface (m)	Percentage by weight in gravel (+4–64 mm) fraction					
	Limestone including chalk	Flint	Ironstone	Sandstone	Quartzite	Others
1.1–4.0	42	12	34	0	12	0

TF 00 NE 67 0740 0984 Near Casewick Hall Lodge, Stamford **Block O**

Surface level (+23.2 m) +76 ft
Water not struck
152 mm percussion
March 1976

Overburden 1.6 m
Mineral 1.2 m
Bedrock 2.5 m+

LOG

Geological classification	Lithology	Thickness m	Depth m
	Soil, brown	0.4	0.4
Glacial Sand and Gravel	Clay, stiff, dark orange-brown to pale yellowish brown, sandy and pebbly	1.2	1.6
	'Very clayey' pebbly sand Gravel: fine with trace coarse, mainly limestone and ironstone Sand: mainly fine with trace amounts of medium and coarse, quartz with limestone ooliths Fines: orange to orange-brown	1.2	2.8
?Kellaways Sand	Clay, stiff, mottled dark grey to rust-brown in upper 1.7 m becoming dark grey, silty; below 4.5 m occasional listric surfaces and traces of mica	2.5+	5.3

GRADING

Mean for deposit *percentages* Fines	Sand	Gravel	Depth below surface (m)	*percentages* Fines $-\frac{1}{16}$	Sand $+\frac{1}{16}-\frac{1}{4}$	$+\frac{1}{4}-1$	$+1-4$	Gravel $+4-16$	$+16-64$	$+64$
25	66	9	1.6–2.0	27	25	14	11	20	3	0
			2.0–2.8	24	69	3	3	1	0	0
			Mean	25	54	6	6	7	2	0

TF 00 SE 37 0791 0188 Near Wittering Lodge, Southorpe **Block O**

Surface level (+16.0 m) +52.5 ft
Water struck at (+13.1 m)
152 mm percussion
March 1976

Overburden 1.9 m
Mineral 4.0 m
Bedrock 1.1 m+

LOG

Geological classification	Lithology	Thickness m	Depth m
	Soil, greyish brown	0.3	0.3
Alluvium	Clay, greyish brown mottled dark ochre-brown, silty, micaceous	0.6	0.9
River Terrace Deposits (First Terrace)	Clay, pale grey mottled orange-brown with fine quartz sand and an occasional cobble of limestone	1.0	1.9
	'Clayey' gravel. 'Very clayey' in upper 1.1 m Gravel: fine with some coarse, mainly rounded to subangular limestone with flint and ironstone and some sandstone and quartzite Sand: fine to coarse, quartz, flint, and limestone Fines: orange-brown	4.0	5.9
Lower Estuarine 'Series'	Clay, stiff, dark grey, silty, pebbly in upper 0.4 m, occasional shell fragments	1.1+	7.0

GRADING

Mean for deposit percentages			Depth below surface (m)	percentages						
Fines	Sand	Gravel		Fines	Sand			Gravel		
				$-\frac{1}{16}$	$+\frac{1}{16}-\frac{1}{4}$	$+\frac{1}{4}-1$	+1–4	+4–16	+16–64	+64
12	40	48	1.9–3.0	36	39	14	3	7	1	0
			3.0–4.1	2	3	5	18	50	22	0
			4.1–5.1	3	5	13	12	46	21	0
			5.1–5.9	5	7	27	12	29	20	0
			Mean	12	14	15	11	33	15	0

COMPOSITION

Depth below surface (m)	Percentage by weight in gravel (+4–16 mm) fraction					
	Limestone including chalk	Flint	Ironstone	Sandstone	Quartzite	Others
1.9–5.9	44	17	23	2	14	0

TF 00 SE 38 0795 0105 Near Sacrewell Lodge, Thornhaugh **Block O**

Surface level (+18.9 m) +62 ft
Water struck at (+16.4 m)
152 mm percussion
February 1976

Overburden 0.6 m
Mineral 6.5 m
Bedrock 1.4 m+

LOG

Geological classification	Lithology	Thickness m	Depth m
	Soil, brown	0.2	0.2
River Terrace Deposits (First Terrace)	Clay, pale orange-brown, silty, occasional pebble	0.4	0.6
	'Clayey' sandy gravel Gravel: fine with trace coarse, mainly shelly and oolitic limestone with angular to subangular brown flint and subrounded to rounded ironstone. Some quartzite Sand: medium and coarse with some fine, subangular quartz with limestone ooliths and ?shell fragments Fines: brown to pale brown	6.5	7.1
Upper Lias	Clay, stiff, dark grey, silty, with listric surfaces and trace shell fragments	1.4+	8.5

GRADING

Mean for deposit *percentages*			Depth below surface (m)	*percentages*						
Fines	Sand	Gravel		Fines	Sand			Gravel		
				$-\frac{1}{16}$	$+\frac{1}{16}-\frac{1}{4}$	$+\frac{1}{4}-1$	+1–4	+4–16	+16–64	+64
10	49	41	0.6–1.6	17	9	23	20	28	3	0
			1.6–2.6	16	15	24	18	23	4	0
			2.6–3.6	6	7	19	20	44	4	0
			3.6–4.6	4	2	6	35	45	8	0
			4.6–5.6	6	8	21	16	39	10	0
			5.6–6.6	6	7	32	8	33	14	0
			6.6–7.1	19	16	23	21	18	3	0
			Mean	10	9	21	19	34	7	0

COMPOSITION

Depth below surface (m)	Percentage by weight in gravel (+4–64 mm) fraction					
	Limestone including chalk	Flint	Ironstone	Sandstone	Quartzite	Others
0.6–7.1	63	18	13	0	6	0

TF 00 SE 39 0850 0024 Near Sacrewell Farm, Sutton Heath **Block O**

Surface level (+16.1 m) +53 ft
Water struck at (+13.4 m)
152 mm percussion
February 1976

Overburden 0.5 m
Mineral 5.4 m
Bedrock 1.1 m+

LOG

Geological classification	Lithology	Thickness m	Depth m
	Soil, brown passes into pale brown clay with trace limestone pebble	0.5	0.5
River Terrace Deposits (First Terrace)	Gravel Gravel: fine with some coarse, mainly subrounded to rounded shelly and oolitic limestone with angular to subangular flint and subrounded ironstone. Some subrounded quartzite Sand: coarse and medium with trace fine, mainly limestone ooliths with flint ironstone and quartz Fines: yellow to orange-brown	5.4	5.9
Upper Lias	Clay, stiff, dark grey, silty, listric surfaces and trace shell fragments throughout	1.1+	7.0

GRADING

Mean for deposit *percentages* Fines	Sand	Gravel	Depth below surface (m)	*percentages* Fines $+\frac{1}{16}$	Sand $+\frac{1}{16}-\frac{1}{4}$	$+\frac{1}{4}-1$	$+1-4$	Gravel $+4-16$	$+16-64$	$+64$
6	45	49	0.5–1.5	5	4	13	24	47	7	0
			1.5–2.5	13	6	14	24	29	15	0
			2.5–3.5	4	4	12	23	45	12	0
			3.5–4.5	3	3	17	21	39	17	0
			4.5–5.5	4	1	8	60	27	0	0
			5.5–5.9	5	5	9	14	37	30	0
			Mean	6	3	13	29	37	12	0

COMPOSITION

Depth below surface (m)	Percentage by weight in gravel (+4–64 mm) fraction: Limestone including chalk	Flint	Ironstone	Sandstone	Quartzite	Others
0.5–5.9	66	13	14	0	5	2

TF 00 SE 40 0921 0494 North of Ufford, Stamford **Sub-block F_1**

Surface level (+18.9 m) +62 ft	Overburden 0.9 m
Water struck at (+17.3 m)	Mineral 3.1 m
152 mm percussion	Bedrock 2.3 m+
February 1976	

LOG

Geological classification	Lithology	Thickness m	Depth m
	Soil, brown passes into brown clay with limestone gravel	0.9	0.9
River Terrace Deposits (First Terrace)	Gravel	3.1	4.0
	Gravel: fine with some coarse, mainly rounded to subrounded oolitic limestone with angular to subangular flint, rounded to subrounded ironstone and some sandstone		
	Sand: medium and coarse with trace fine, quartz, limestone, ooliths, ironstone and flint		
	Fines: orange-brown		
Oxford Clay	Clay, brown to grey, silty with many weathered limestone pebbles	0.6	4.6
Kellaways Sand	Silt, firm, dark grey, occasional shell fragments	1.7+	6.3

GRADING

Mean for deposit *percentages* Fines	Sand	Gravel	Depth below surface (m)	*percentages* Fines	Sand			Gravel		
				$-\frac{1}{16}$	$+\frac{1}{16}-\frac{1}{4}$	$+\frac{1}{4}-1$	+1–4	+4–16	+16–64	+64
6	46	48	0.9–1.9	10	10	26	13	35	6	0
			1.9–2.9	3	5	21	21	40	10	0
			2.9–4.0	4	7	20	16	34	19	0
			Mean	6	7	22	17	36	12	0

COMPOSITION

Depth below surface (m)	Percentage by weight in gravel (+4–64 mm) fraction: Limestone including chalk	Flint	Ironstone	Sandstone	Quartzite	Others
0.9–4.0	60	22	12	5	1	0

TF 10 NW 28 **1097 0952** **Windmill Field, near West Deeping** **Sub-block F_2**

Surface level (+11.6 m) +38 ft
Water struck at (+9.4 m)
152 mm percussion
February 1976

Overburden 1.0 m
Mineral 5.3 m
Bedrock 1.0 m+

LOG

Geological classification	Lithology	Thickness m	Depth m
	Soil, yellowish brown, stony, with an indurated clay layer below 0.9 m	1.0	1.0
River Terrace Deposits (First Terrace)	Gravel Gravel: mainly fine with some coarse and trace cobble, mostly rounded to subangular shelly and oolitic limestone with flint, ironstone and quartzite Sand: coarse and medium, with trace fine, quartz and limestone ooliths Fines: pale brown	5.3	6.3
Oxford Clay	Clay, indurated, greyish green becoming blue towards base, with calcareous siltstone nodules and bivalve shell fragments	1.0+	7.3

GRADING

Mean for deposit *percentages* Fines	Sand	Gravel	Depth below surface (m)	*percentages* Fines $-\frac{1}{16}$	Sand $+\frac{1}{16}-\frac{1}{4}$	$+\frac{1}{4}-1$	$+1-4$	Gravel $+4-16$	$+16-64$	$+64$
5	44	51	1.0–2.0	8	10	23	14	41	5	0
			2.0–2.5	8	7	26	17	36	6	0
			2.5–3.5	3	5	20	19	44	9	0
			3.5–4.5	6	5	12	24	48	6	0
			4.5–5.5	4	5	21	20	28	13	11
			5.5–6.3	4	3	15	20	36	13	9
			Mean	5	6	19	19	39	9	3

COMPOSITION

Depth below surface (m)	Percentage by weight in gravel (+4–64 m) fraction: Limestone including chalk	Flint	Ironstone	Sandstone	Quartzite	Others
1.0–6.3	72	17	9	0	2	0

TF 10 NW 29 1016 0860 Horse Marsh, near West Deeping **Sub-block A_1**

Surface level (+12.8 m)+42 ft
Water struck at (+10.4 m)
152 mm percussion
February 1976

Overburden 2.4 m
Mineral 2.6 m
Bedrock 1.0 m+

LOG

Geological classification	Lithology	Thickness m	Depth m
Made ground		1.1	1.1
Alluvium with ?Lower Peat	Clay, firm, mottled orange and grey, becomes dark brown and peaty below 1.9 m	1.0	2.1
River Terrace Deposits (First Terrace)	Silt, very soft, grey, with sandy layers	0.3	2.4
	Gravel Gravel: fine with some coarse and trace cobble, subangular to rounded oolitic and shelly limestone with angular flint and subrounded ironstone with some quartzite and sandstone Sand: coarse and medium and trace fine Fines: buff-yellow to yellow	2.6	5.0
?Oxford Clay	Clay, with black limestone layer in upper 0.1 m, dark grey, silty, with bivalve shells and belemnite fragments	0.5	5.5
?Kellaways Clay	Siltstone, friable, greenish grey, with small calcareous nodules	0.5+	6.0

GRADING

Mean for deposit *percentages* Fines	Sand	Gravel	Depth below surface (m)	*percentages* Fines	Sand			Gravel		
				$-\frac{1}{16}$	$+\frac{1}{16}-\frac{1}{4}$	$+\frac{1}{4}-1$	+1–4	+4–16	+16–64	+64
3	45	52	2.4–3.4	3	4	21	25	39	8	1
			3.4–4.4	4	2	10	34	42	7	1
			4.4–5.0	2	1	5	29	50	12	0
			Mean	3	3	13	29	43	8	1

COMPOSITION

Depth below surface (m)	Percentage by weight in gravel (+4–64 mm) fraction: Limestone including chalk	Flint	Ironstone	Sandstone	Quartzite	Others
2.4–5.0	59	17	16	1	6	1

TF 10 NW 31 **1035 0630** **South of Helpston Ballast Pit, Bainton** **Sub-block F_1**

Surface level (+13.7 m) +45 ft
Water struck at (+11.5 m)
152 mm percussion
February 1976

Overburden 0.8 m
Mineral 5.5 m
Bedrock 0.7 m+

LOG

Geological classification	Lithology	Thickness m	Depth m
	Soil, dark brown to black, sandy, trace gravel	0.3	0.3
River Terrace Deposits (First Terrace)	Clay, pale brown, sandy, ironstained with pebbles of brown flint and ironstone	0.5	0.8
	Gravel Gravel: fine with some coarse and trace cobble, mainly black to white angular flint, with subrounded to rounded oolitic limestone and some ironstone Sand: medium and coarse with trace fine, quartz, flint, ironstone with limestone ooliths Fines: pale brown to buff	5.5	6.3
Kellaways Clay	Clay, indurated, grey, slightly silty	0.7+	7.0

GRADING

Mean for deposit *percentages* Fines	Sand	Gravel	Depth below surface (m)	*percentages* Fines $-\frac{1}{16}$	Sand $+\frac{1}{16}-\frac{1}{4}$	$+\frac{1}{4}-1$	$+1-4$	Gravel $+4-16$	$+16-64$	$+64$
6	46	48	0.8–2.2	8	12	25	16	36	4	0
			2.2–3.2	5	5	21	22	39	7	0
			3.2–4.2	4	6	26	15	38	6	6
			4.2–5.3	6	2	15	22	39	15	2
			5.3–6.3	7	4	19	20	33	12	6
			Mean	6	6	21	19	37	8	3

COMPOSITION

Depth below surface (m)	Percentage by weight in gravel (+4–64 mm) fraction: Limestone including chalk	Flint	Ironstone	Sandstone	Quartzite	Others
0.8–6.3	21	71	7	0	0	1

TF 10 NW 32 1164 0889 West Deeping Sub-block A_1

Surface level (+10.3 m) +34 ft
Water struck at (+8.3 m)
152 mm percussion
February 1976

Overburden 1.6 m
Mineral 3.4 m
Waste 0.5 m
Mineral 0.8 m
Bedrock 0.8 m+

LOG

Geological classification	Lithology	Thickness m	Depth m
	Soil, brown	0.5	0.5
River Terrace Deposits (First Terrace)	Clay, firm to indurated, brown to orange, sandy patches, some limestone and flint pebbles	1.1	1.6
	a Gravel Gravel: fine with some coarse and trace cobble, mainly rounded quartzite and shelly and oolitic limestone with some angular to rounded flint and subrounded ironstone Sand: coarse and medium with trace fine Fines: yellowish brown	3.4	5.0
	Clay, soft, orange becoming greenish buff to blue	0.5	5.5
	b Gravel—as above	0.8	6.3
?Kellaways Sand	No information available	0.8+	7.1

GRADING

	Mean for deposit *percentages* Fines	Sand	Gravel	Depth below surface (m)	*percentages* Fines $-\frac{1}{16}$	Sand $+\frac{1}{16}-\frac{1}{4}$	$+\frac{1}{4}-1$	$+1-4$	Gravel $+4-16$	$+16-64$	$+64$
a	4	44	52	1.6–2.2	11	6	25	20	32	7	0
				2.2–3.2	3	2	10	27	57	7	1
				3.2–4.2	2	2	14	24	49	8	1
				4.2–5.0	3	5	21	23	40	8	1
				Mean	4	4	16	24	44	7	1
				5.0–5.5 waste							
b	5	45	50	5.5–6.3	5	6	19	21	43	8	0
a+b	4	44	52	Mean	4	4	17	23	44	7	1

COMPOSITION (a+b)

Depth below surface (m)	Percentage by weight in gravel (+4–64 mm) fraction: Limestone including chalk	Flint	Ironstone	Sandstone	Quartzite	Others
1.6–6.3	35	12	12	0	40	1

TF 10 NW 33 1105 0830 Church Meadow, West Deeping **Sub-block A_1**

Surface level (+11.6 m) +38 ft
Water struck at (+9.3 m)
152 mm percussion
February 1976

Overburden 2.3 m
Mineral 1.2 m
Bedrock 0.8 m+

LOG

Geological classification	Lithology	Thickness m	Depth m
	Soil, brown, clayey	0.1	0.1
Alluvium	Clay, stiff, yellow to dark grey with pale grey mottling in upper part, becomes silty below 1.9 m	2.2	2.3
River Terrace Deposits (First Terrace)	Gravel Gravel: fine with coarse, mainly rounded to subangular oolitic and crystalline limestone, with sandstone, ironstone, flint and quartzite Sand: coarse and medium with trace fine Fines: dark grey	1.2	3.5
Kellaways Sand	Siltstone, 'clayey' in upper 0.1 m, greyish green with dark brown to purple patches, bivalve shell and belemnite fragments present	0.8+	4.3

GRADING

Mean for deposit *percentages* Fines	Sand	Gravel	Depth below surface (m)	*percentages* Fines $-\frac{1}{16}$	Sand $+\frac{1}{16}-\frac{1}{4}$	$+\frac{1}{4}-1$	$+1-4$	Gravel $+4-16$	$+16-64$	$+64$
4	38	58	2.3–3.3	4	2	17	19	39	19	0
			3.3–3.5	4	3	16	20	38	19	0
			Mean	4	3	16	19	39	19	0

COMPOSITION

Depth below surface (m)	Percentage by weight in gravel (+4–64 mm) fraction: Limestone including chalk	Flint	Ironstone	Sandstone	Quartzite	Others
2.3–3.5	41	7	17	33	2	0

TF 10 NW 34 1191 0798 Opposite St. Peter's Church, Maxey **Sub-block F_1**

Surface level (+11.0 m) +36 ft
Water struck at (+8.9 m)
152 mm percussion
February 1976

Overburden 0.5 m
Mineral 4.4 m
Waste 0.7 m
Mineral 0.4 m
Bedrock 1.5 m+

LOG

Geological classification	Lithology	Thickness m	Depth m
	Soil, brown	0.3	0.3
River Terrace Deposits (First Terrace)	Clay, pale brown mottled orange-brown, silty, trace flint pebbles	0.2	0.5
	a Gravel Gravel: mainly fine with some coarse and trace cobble especially towards base, mostly subrounded to rounded, tabular, shelly and oolitic limestone with brown to dark reddish brown ironstone, flint, sandstone and some quartzite Sand: coarse and medium with trace fine, quartz and ironstone with some limestone ooliths Fines: pale greyish brown to brown	4.4	4.9
	Clay, soft, dark khaki-grey, silty, glutinous	0.7	5.6
	b Gravel—as above	0.4	6.0
?Oxford Clay	Clay stiff, dark grey, silty, trace shell fragments	0.9	6.9
?Kellaways Sand	'Clayey' sand, dark grey, silty, with coarse sand	0.6+	7.5

GRADING

	Mean for deposit *percentages*			Depth below surface (m)	*percentages*						
	Fines	Sand	Gravel		Fines	Sand			Gravel		
					$-\frac{1}{16}$	$+\frac{1}{16}-\frac{1}{4}$	$+\frac{1}{4}-1$	+1–4	+4–16	+16–64	+64
a	3	44	53	0.5–1.5	5	6	30	20	34	5	0
				1.5–2.5	2	3	10	18	45	12	0
				2.5–3.5	3	3	17	20	44	12	1
				3.5–4.5	3	3	16	24	45	9	0
				4.5–4.9	3	3	15	20	38	13	8
				Mean	3	3	20	21	42	10	1
				4.9–5.6 waste							
b	9	36	55	5.6–6.0	9	3	8	25	45	7	3
a+b	4	43	53	Mean	4	3	19	21	42	10	1

COMPOSITION (a+b)

Depth below surface (m)	Percentage by weight in gravel (+4–64 mm) fraction					
	Limestone including chalk	Flint	Ironstone	Sandstone	Quartzite	Others
0.5–6.0	55	11	17	14	2	1

TF 10 NW 35 **1116 0654** **South of Lolham Bridges, Bainton** **Sub-block F_1**

Surface level (+12.2 m) +40 ft
Water struck at (+9.6 m)
152 mm percussion
March 1976

Overburden 0.2 m
Mineral 5.0 m
Bedrock 0.4 m+

LOG

Geological classification	Lithology	Thickness m	Depth m
	Soil, brown sandy	0.2	0.2
River Terrace Deposits (First Terrace)	'Clayey' sandy gravel Gravel: fine with trace coarse, traces of cobble between 3.2–4.2 m, mainly subrounded to rounded oolitic limestone with sandy ironstone, subangular to angular grey flint, rounded brown sandstone and some quartzite Sand: medium and coarse with some fine, quartz with limestone ooliths Fines: pale brown	5.0	5.2
Cornbrash	Limestone, shelly, bluish grey	0.4+	5.6

GRADING

Mean for deposit percentages			Depth below surface (m)	percentages						
Fines	Sand	Gravel		Fines	Sand			Gravel		
				$-\frac{1}{16}$	$+\frac{1}{16}-\frac{1}{4}$	$+\frac{1}{4}-1$	+1–4	+4–16	+16–64	+64
12	51	37	0.2–1.2	32	23	27	7	11	2	0
			1.2–2.2	7	9	34	14	30	6	0
			2.2–3.2	13	8	32	9	33	4	0
			3.2–4.2	3	3	18	23	42	9	2
			4.2–5.2	4	4	16	28	40	9	0
			Mean	12	10	25	16	31	6	0

COMPOSITION

Depth below surface (m)	Percentage by weight in gravel (+4–64 mm) fraction					
	Limestone including chalk	Flint	Ironstone	Sandstone	Quartzite	Others
0.2–5.2	38	16	25	15	6	0

TF 10 NW 36 1159 0942 Lammas Close, West Deeping **Sub-block F_2**

Surface level (+10.0 m) +33 ft
Water struck at (+7.6 m)
152 mm percussion
February 1976

Overburden 0.9 m
Mineral 5.4 m
Bedrock 1.0 m+

LOG

Geological classification	Lithology	Thickness m	Depth m
	Soil, brown	0.4	0.4
River Terrace Deposits (First Terrace)	Clay, pale brown to orange-brown, silty trace flint pebbles	0.5	0.9
	Gravel	5.4	6.3
	Gravel: fine with some coarse and trace cobble, mainly subrounded, tabular, oolitic and shelly limestone with ironstone and angular to subangular brown flint		
	Sand: medium and coarse with trace fine, quartz, limestone ooliths and flint		
	Fines: pale brown to pale greyish brown		
Oxford Clay	Clay, dark grey, silty, with shell fragments including belemnites	1.0+	7.3

GRADING

Mean for deposit percentages			Depth below surface (m)	percentages						
Fines	Sand	Gravel		Fines	Sand			Gravel		
				$-\frac{1}{16}$	$+\frac{1}{16}-\frac{1}{4}$	$+\frac{1}{4}-1$	+1–4	+4–16	+16–64	+64
4	45	51	0.9–1.9	5	6	24	15	43	7	0
			1.9–2.9	2	5	20	17	45	9	1
			2.9–3.9	3	4	19	21	44	8	1
			3.9–4.9	4	4	24	18	37	8	5
			4.9–6.3	3	5	30	13	35	11	3
			Mean	4	5	24	16	40	9	2

COMPOSITION

Depth below surface (m)	Percentage by weight in gravel (+4–64 mm) fraction					
	Limestone including chalk	Flint	Ironstone	Sandstone	Quartzite	Others
0.9–6.3	58	13	18	8	2	1

TF 10 NW 37 **1275 0883** **Castle Farm, Maxey** **Sub-block A_1**

Surface level (+8.8 m) +29 ft
Water level not recorded
152 mm percussion
February 1976

Overburden 1.1 m
Mineral 6.0 m
Bedrock 1.4 m+

LOG

Geological classification	Lithology	Thickness m	Depth m
	Soil, dark brown	0.2	0.2
Alluvium	Clay, pale brown, mottled with reddish-brown, trace flint gravel	0.9	1.1
River Terrace Deposits (First Terrace)	Gravel Gravel: fine with some coarse and trace cobble, mainly subrounded to rounded, shelly and oolitic limestone with brown subangular to subrounded flint and some subrounded black ironstone with occasional quartzites and sandstones Sand: medium and coarse with trace fine, mainly ironstone with flint, limestone ooliths and sandstone Fines: pale greyish brown	6.0	7.1
Kellaways Clay	Clay, stiff, dark greyish brown, silty, with abudant shell fragments below 7.9 m	1.4+	8.5

GRADING

Mean for deposit percentages			Depth below surface (m)	percentages						
Fines	Sand	Gravel		Fines	Sand			Gravel		
				$-\frac{1}{16}$	$+\frac{1}{16}-\frac{1}{4}$	$+\frac{1}{4}-1$	+1–4	+4–16	+16–64	+64
4	45	51	1.1–2.1	8	9	29	16	35	4	0
			2.1–3.1	3	4	22	20	41	10	0
			3.1–4.1	3	3	26	19	39	10	1
			4.1–5.1	3	4	20	17	44	10	3
			5.1–6.1	2	3	19	21	35	15	5
			6.1–7.1	3	4	19	19	35	12	8
			Mean	4	4	23	18	38	10	3

COMPOSITION

Depth below surface (m)	Percentage by weight in gravel (+4–64 mm) fraction					
	Limestone including chalk	Flint	Ironstone	Sandstone	Quartzite	Others
1.1–7.1	48	25	21	2	4	0

TF 10 NW 38 1216 0731 Nunton House Farm, Maxey **Sub-block F_1**

Surface level (+9.7 m) +32 ft
Water struck at (+7.2 m)
152 mm percussion
February 1976

Overburden 0.4 m
Mineral 5.1 m
Bedrock 1.1 m+

LOG

Geological classification	Lithology	Thickness m	Depth m
	Soil	0.3	0.3
River Terrace Deposits (First Terrace)	Clay, brown, sandy, with some gravel	0.1	0.4
	Gravel	5.1	5.5
	Gravel: fine with some coarse and trace cobble, mainly rounded to subrounded, tabular, limestone with dark reddish brown ironstone, angular to subangular brown flint and quartzite		
	Sand: coarse and medium with trace fine, mainly quartz with ironstone, with some flint and occasional limestone ooliths		
	Fines: pale orange-brown to brown		
Kellaways Clay	Clay, very stiff, dark grey, trace shell fragment	1.1+	6.6

GRADING

Mean for deposit percentages			Depth below surface (m)	percentages						
Fines	Sand	Gravel		Fines	Sand			Gravel		
				$-\frac{1}{16}$	$+\frac{1}{16}-\frac{1}{4}$	$+\frac{1}{4}-1$	+1–4	+4–16	+16–64	+64
4	43	53	0.4–1.4	8	6	25	18	36	7	0
			1.4–2.4	2	3	17	20	47	12	0
			2.4–3.4	2	3	20	20	44	10	1
			3.4–4.4	2	3	19	19	46	10	1
			4.4–5.5	5	4	18	21	43	7	1
			Mean	4	4	20	19	43	9	1

COMPOSITION

Depth below surface (m)	Percentage by weight in gravel (+4–64 mm) fraction					
	Limestone including chalk	Flint	Ironstone	Sandstone	Quartzite	Others
0.4–5.5	53	18	25	0	3	1

TF 10 NW 39 **1319 0992** **Market Deeping** **Sub-block A_1**

Surface level (+1.2 m) +4 ft
Water struck at O.D.
152 mm percussion
February 1976

Overburden 0.9 m
Mineral 3.1 m
Bedrock 1.2 m+

LOG

Geological classification	Lithology	Thickness m	Depth m
	Soil, dark greyish brown	0.2	0.2
Alluvium	Clay, brown, silty, slightly mottled with rust patches, trace flint gravel, many small white rootlets	0.7	0.9
River Terrace Deposits (First Terrace)	Gravel Gravel: fine with some coarse and trace cobble, mainly rounded to subrounded crystalline, shelly and oolitic limestone with angular to subangular flint and some ironstone, sandstone and quartzite, trace fossil fragments Sand: coarse and medium with trace fine, mainly quartz with ironstone and some limestone ooliths Fines: pale brown to grey	3.1	4.0
Kellaways Sand	Siltstone, alternations of well indurated and softer layers, dark grey	1.2+	5.2

GRADING

Mean for deposit *percentages*			Depth below surface (m)	*percentages*						
Fines	Sand	Gravel		Fines	Sand			Gravel		
				$-\frac{1}{16}$	$+\frac{1}{16}-\frac{1}{4}$	$+\frac{1}{4}-1$	+1–4	+4–16	+16–64	+64
3	38	59	0.9–1.9	5	4	18	18	40	10	5
			1.9–2.9	3	1	15	26	41	11	3
			2.9–4.0	1	1	11	19	58	10	0
			Mean	3	2	15	21	47	10	2

COMPOSITION

Depth below surface (m)	Percentage by weight in gravel (+14–64 mm) fraction					
	Limestone including chalk	Flint	Ironstone	Sandstone	Quartzite	Others
0.9–4.0	76	11	5	4	4	0

TF 10 NW 40 **1346 0912** **Foxcovert Barn, Maxey** **Sub-block A_1**

Surface level (+7.6 m) +25 ft
Water struck at (+5.6 m)
152 mm percussion
February 1976

Overburden 1.0 m
Mineral 2.2 m
Bedrock 0.5 m+

LOG

Geological classification	Lithology	Thickness m	Depth m
Made ground	Soil, dark brown, passes into clay with gravel and brick rubble	0.3	0.3
Alluvium	Clay, pale brown to yellowish brown, silty, trace flint gravel	0.7	1.0
River Terrace Deposits (First Terrace)	Sandy gravel Gravel: fine with trace coarse, mainly subrounded, tabular, oolitic limestone with black to brown ironstone, quartzite with subangular brown flint Sand: medium and coarse with trace fine, mainly quartz with limestone ooliths and ironstone Fines: pale yellowish brown to greyish brown	2.2	3.2
Kellaways Sand	Siltstone, well indurated, dark grey	0.5+	3.7

GRADING

Mean for deposit percentages			Depth below surface (m)	*percentages*						
Fines	Sand	Gravel		Fines	Sand			Gravel		
				$-\frac{1}{16}$	$+\frac{1}{16}-\frac{1}{4}$	$+\frac{1}{4}-1$	+1–4	+4–16	+16–64	+64
3	51	46	1.0–2.0	5	10	28	19	34	5	0
			2.0–3.2	4	3	21	21	44	7	0
			Mean	3	7	24	20	40	6	0

COMPOSITION

Depth below surface (m)	Percentage by weight in gravel (+4–64 mm) fraction					
	Limestone including chalk	Flint	Ironstone	Sandstone	Quartzite	Others
1.0–3.2	55	9	18	0	18	0

TF 10 NW 41 **1305 0793** **Maxey** **Sub-block A_1**

Surface level (+9.4 m) +31 ft
Water struck at (+6.7 m)
152 mm percussion
February 1976

Overburden 0.9 m
Mineral 6.1 m
Bedrock 0.6 m+

LOG

Geological classification	Lithology	Thickness m	Depth m
	Soil, dark brown	0.3	0.3
Alluvium	Clay, brown mottled in places dark red, silty, trace flint gravel at base	0.6	0.9
River Terraces Deposits (First Terrace)	Gravel Gravel: fine with some coarse and trace cobble, mainly rounded to subrounded, oolitic and shelly limestone with subrounded brown ironstone, angular to subangular flint and some quartzite Sand: coarse and medium with trace fine, mainly quartz with limestone ooliths, ironstone, flint and quartzite Fines, pale brown	6.1	7.0
Kellaways Clay	Clay, friable, dark grey, silty, with shell fragments	0.6	7.6

GRADING

Mean for deposit percentages Fines	Sand	Gravel	Depth below surface (m)	percentages Fines $-\frac{1}{16}$	Sand $+\frac{1}{16}-\frac{1}{4}$	$+\frac{1}{4}-1$	$+1-4$	Gravel $+4-16$	$+16-64$	$+64$
4	42	54	0.9–1.9	7	6	25	18	40	4	0
			1.9–2.9	3	3	20	22	43	9	1
			2.9–3.9	3	1	12	24	49	11	0
			3.9–4.9	3	2	12	29	43	7	5
			4.9–5.9	3	1	15	23	46	11	1
			5.9–7.0	3	2	15	23	41	13	1
			Mean	4	2	17	23	44	9	1

COMPOSITION

Depth below surface (m)	Percentages by weight in gravel (+4–64 mm) fraction: Limestone including chalk	Flint	Ironstone	Sandstone	Quartzite	Others
0.9–7.0	58	11	25	1	4	1

TF 10 NW 42 **1317 0696** **Woodgate Lane, Maxey** **Sub-block A_1**

Surface level (+8.5 m) +28 ft
Water struck at (+6.2 m)
152 mm percussion
February 1976

Overburden 0.6 m
Mineral 3.1 m
Waste 0.8 m
Mineral 0.5 m
Bedrock 0.9 m+

LOG

Geological classification	Lithology	Thickness m	Depth m
	Soil, dark brown	0.1	0.1
Alluvium	Clay, soft to firm, pale orange-brown, slightly mottled reddish brown, silty with trace gravel	0.5	0.6
River Terrace Deposits (First Terrace)	**a** Gravel Gravel: fine with trace coarse, trace cobble between 0.6–1.6 m, mainly rounded to subrounded, tabular limestone with black to dark brown ironstone, with flint, some quartzite and traces of sandstone Sand: medium with coarse and trace fine, quartz, limestone ooliths, ironstone and flint Fines: pale brown to pale orange-brown	3.1	3.7
	Silt, soft to firm, dark grey to black speckled in places with orange, many black ?carbonaceous patches, micaceous in part	0.8	4.5
	b Sandy gravel Gravel: fine with some coarse and trace cobble, mainly limestone with ironstone, flint, quartzite and sandstone Sand: medium with some fine and trace coarse Fines: pale brown	0.5	5.0
Kellaways Clay	Clay, dark grey to dark olive-grey, silty, trace shell fragments and carbonaceous material	0.9+	5.9

GRADING

	Mean for deposit *percentages*			Depth below surface (m)	*percentages*						
	Fines	Sand	Gravel		Fines	Sand			Gravel		
					$-\frac{1}{16}$	$+\frac{1}{16}-\frac{1}{4}$	$+\frac{1}{4}-1$	+1–4	+4–16	+16–64	+64
a	6	46	48	0.6–1.6	6	4	26	21	36	7	1
				1.6–2.6	3	5	24	20	43	5	0
				2.6–3.7	8	2	19	17	44	8	0
				Mean	6	4	23	19	41	7	0
				3.7–4.5 waste							
b	8	60	32	4.5–5.0	8	15	38	7	21	9	2
a+b	6	48	46	Mean	6	5	25	18	38	7	1

COMPOSITION (a+b)

Depth below surface (m)	Percentage by weight in gravel (+4–64 mm) fraction					
	Limestone including chalk	Flint	Ironstone	Sandstone	Quartzite	Others
0.6–5.0	46	23	23	1	6	1

TF 10 NW 43 **1310 0583** **Old Railway Hotel, Helpston railway crossing** **Sub-block F_1**

Surface level (+10.0 m) +33 ft
Water struck at (+7.7 m)
152 mm percussion
February 1976

Overburden 1.0 m
Mineral 1.8 m
Bedrock 0.9 m+

LOG

Geological classification	Lithology	Thickness m	Depth m
Made ground	Soil, with brick and gravel rubble	0.2	0.2
River Terrace Deposits (First Terrace)	Clay, firm to stiff, pale brown, silty with coarse gravel towards base	0.8	1.0
	Sandy gravel Gravel: fine with trace coarse, mainly rounded to subrounded, tabular, shelly and oolitic limestone with ironstone and angular to subangular flint Sand: medium with coarse and some fine, mainly flint and quartz with ironstone Fines: orange-brown	1.8	2.8
?Blisworth Clay	Clay, soft to firm, dark grey to black, silty, trace shell fragments, passes into a fossiliferous siltstone at base	0.9+	3.7

GRADING

Mean for deposit *percentages*			Depth below surface (m)	*percentages*						
Fines	Sand	Gravel		Fines	Sand			Gravel		
				$-\frac{1}{16}$	$+\frac{1}{16}-\frac{1}{4}$	$+\frac{1}{4}-1$	+1–4	+4–16	+16–64	+64
5	58	37	1.0–2.0	4	9	31	14	37	5	0
			2.0–2.8	6	7	35	21	28	3	0
			Mean	5	8	33	17	33	4	0

COMPOSITION

Depth below surface (m)	Percentage by weight in gravel (+4–64 mm) fraction					
	Limestone including chalk	Flint	Ironstone	Sandstone	Quartzite	Others
1.0–2.8	46	15	21	3	15	0

TF 10 NW 44 1402 0863 Foxcovert Barn Farm, Maxey **Sub-block F_3**

Surface level (+7.3 m) +24 ft
Water level not recorded
152 mm percussion
February 1976

Overburden 0.6 m
Mineral 4.3 m
Bedrock 1.6 m+

LOG

Geological classification	Lithology	Thickness m	Depth m
	Soil, brown	0.3	0.3
River Terrace Deposits (First Terrace)	Clay, pale brown, firm, trace gravel	0.3	0.6
	Gravel, iron-pan layer in upper 0.9 m Gravel: fine with some coarse and trace cobble, mainly subrounded tabular, shelly and oolitic limestone and rounded to subrounded ironstone with flint, quartzite and sandstone Sand: coarse and medium with trace fine, quartz, ironstone and limestone Fines: pale greyish brown	4.3	4.9
Kellaways Clay	Clay, stiff, grey, silty	1.6+	6.5

GRADING

Mean for deposit *percentages* Fines	Sand	Gravel	Depth below surface (m)	*percentages* Fines $-\frac{1}{16}$	Sand $+\frac{1}{16}-\frac{1}{4}$	$+\frac{1}{4}-1$	$+1-4$	Gravel $+4-16$	$+16-64$	$+64$
4	39	57	0.6–1.6	8	8	22	19	38	5	0
			1.6–2.6	3	4	18	23	40	12	0
			2.6–3.6	2	3	18	19	43	1	5
			3.6–4.9	2	1	9	17	47	20	3
			Mean	4	4	16	19	42	13	2

COMPOSITION

Depth below surface (m)	Percentage by weight in gravel (+4–64 mm) fraction: Limestone including chalk	Flint	Ironstone	Sandstone	Quartzite	Others
0.6–4.9	49	11	22	6	11	1

TF 10 NW 45 **1397 0785** **Near Maxey** **Sub-block A_1**

Surface level (+7.6 m) +25 ft
Water struck at (+5.8 m)
152 mm percussion
February 1976

Overburden 1.0 m
Mineral 5.4 m
Bedrock 1.3 m+

LOG

Geological classification	Lithology	Thickness m	Depth m
	Soil, dark brown	0.3	0.3
Alluvium	Clay, firm, pale brown, partially mottled reddish brown, silty, with trace gravel	0.7	1.0
River Terrace Deposits (First Terrace)	Gravel, upper 0.2 m mainly sand Gravel: fine with some coarse and trace cobble, mainly angular to subangular, brown flint with subrounded ironstone and some oolitic and shelly limestones Sand: coarse with medium and trace fine, mainly quartz with some flint and ironstone Fines: pale greyish brown	5.4	6.4
Kellaways Clay	Clay, pebbly in upper 0.3 m, stiff, dark grey, silty, ?micaceous below 6.7 m, mainly comminuted shell fragments near base	1.3+	7.7

GRADING

Mean for deposit *percentages* Fines	Sand	Gravel	Depth below surface (m)	*percentages* Fines $-\frac{1}{16}$	Sand $+\frac{1}{16}-\frac{1}{4}$	$+\frac{1}{4}-1$	$+1-4$	Gravel $+4-16$	$+16-64$	$+64$
2	31	67	1.0–2.0	5	4	15	18	53	5	0
			2.0–3.0	1	1	9	16	5	17	5
			3.0–4.0	2	2	11	16	48	14	7
			4.0–5.0	1	2	18	17	48	14	0
			5.0–6.4	2	3	11	13	45	22	6
			Mean	2	2	13	16	48	15	4

TF 10 NW 46 **1406 0674** **Near St. Stephen's Church, Etton** **Sub-block F_1**

Surface level (+7.3 m) +24 ft
Water struck at (+5.3 m)
152 mm percussion
February 1976

Overburden 0.9 m
Mineral 3.2 m
Bedrock 0.5 m+

LOG

Geological classification	Lithology	Thickness m	Depth m
	Soil, dark greyish brown	0.5	0.5
River Terrace Deposits (First Terrace)	Clay, well indurated, orange-brown to pale yellow, with iron-pan layers and flint gravel	0.4	0.9
	Gravel Gravel: fine with some coarse, mainly subrounded brown flint with some oolitic and shelly limestone with sandstone, quartzite and ironstone Sand: medium and coarse with trace fine, flint, ironstone, sandstone and quartz Fines: pale brown	3.2	4.1
?Kellaways Sand	Limestone, dark grey, finely crystalline, with trace *Gryphaea* and belemnite, passes into a firm, dark grey silt below 4.4 m	0.5+	4.6

GRADING

Mean for deposit *percentages*			Depth below surface (m)	*percentages*						
Fines	Sand	Gravel		Fines	Sand			Gravel		
				$-\frac{1}{16}$	$+\frac{1}{16}-\frac{1}{4}$	$+\frac{1}{4}-1$	+1–4	+4–16	+16–64	+64
6	39	55	0.9–1.9	11	5	31	16	34	3	0
			1.9–3.9	3	2	15	18	52	10	0
			3.9–4.1	3	3	10	14	43	7	0
			Mean	6	3	19	17	46	9	0

COMPOSITION

Depth below surface (m)	Percentage by weight in gravel (+4–64 mm) fraction					
	Limestone including chalk	Flint	Ironstone	Sandstone	Quartzite	Others
0.9–4.1	16	69	4	6	5	0

TF 10 NW 47 1396 0589 South of Etton **Sub-block F_1**

Surface level (+7.9 m) +26 ft
Water struck at (+6.3 m)
152 mm percussion
February 1976

Overburden 0.5 m
Mineral 2.2 m
Bedrock 1.1 m+

LOG

Geological classification	Lithology	Thickness m	Depth m
	Soil, brown	0.3	0.3
River Terrace Deposits	Clay, stiff, pale brown, silty with some flint gravel	0.2	0.5
(First Terrace)	Sandy gravel Gravel: fine with trace coarse, mainly subangular to subrounded brown and white flint with sandstone, quartzite, limestone and ironstone Sand: medium and coarse with some fine, quartz, flint and ironstone Fines: pale orange-brown	2.2	2.7
Kellaways Clay	Clay, stiff, grey, variegated in parts, silty, well indurated at base ?limestone	1.1+	3.8

GRADING

Mean for deposit *percentages*			Depth below surface (m)	*percentages*						
Fines	Sand	Gravel		Fines	Sand			Gravel		
				$-\frac{1}{16}$	$+\frac{1}{16}-\frac{1}{4}$	$+\frac{1}{4}-1$	+1–4	+4–16	+16–64	+64
9	52	39	0.5–1.5	12	10	25	20	31	2	0
			1.5–2.7	7	7	23	19	40	4	0
			Mean	9	9	23	20	36	3	0

Surface level (+4.9 m) +16 ft
Water struck at (+2.4 m)
152 mm percussion
February 1976

Overburden 1.4 m
Mineral 2.5 m
Bedrock 2.1 m+

LOG

Geological classification	Lithology	Thickness m	Depth m
	Soil, brown	0.3	0.3
River Terrace Deposits (First Terrace)	Clay, pale orange-brown, silty with trace flint gravel	1.1	1.4
	Gravel Gravel: fine with some coarse and trace cobble, mainly angular to subangular, shelly and oolitic limestone with rounded to subrounded ironstone with some flint and quartzite Sand: medium and coarse with trace fine, mainly quartz with some ironstone, flint and with limestone ooliths Fines: pale orange-brown to greyish brown	2.5	3.9
Kellaways Sand	Clay, firm, greenish grey, silty; passes into dark to pale grey, very fine slightly micaceous sand below 5.0 m	2.1+	6.0

GRADING

Mean for deposit *percentages*			Depth below surface (m)	*percentages*						
Fines	Sand	Gravel		Fines	Sand			Gravel		
				$-\frac{1}{16}$	$+\frac{1}{16}-\frac{1}{4}$	$+\frac{1}{4}-1$	+1–4	+4–16	+16–64	+64
5	45	50	1.4–2.4	5	5	19	17	39	13	1
			2.4–3.4	6	5	20	17	41	11	1
			3.4–3.9	6	8	32	18	31	6	0
			Mean	5	6	22	17	38	11	1

COMPOSITION

Depth below surface (m)	Percentage by weight in gravel (+4–64 mm) fraction					
	Limestone including chalk	Flint	Ironstone	Sandstone	Quartzite	Others
1.4–3.9	41	17	34	0	6	2

TF 10 NE 25 **1509 0870** **North Fen House, near Deeping St. James** **Sub-block F_3**

Surface level (+5.8 m) +19 ft
Water struck at (+4.3)
152 mm percussion
February 1976

Overburden 1.0 m
Mineral 1.9 m
Bedrock 0.8 m+

LOG

Geological classification	Lithology	Thickness m	Depth m
	Soil, brown	0.2	0.2
River Terrace Deposits	Clay, firm, brown, sandy towards base	0.8	1.0
(First Terrace)	Sandy gravel Gravel: fine with trace coarse, mainly limestone with ironstone, subangular to rounded flint and quartzite Sand: medium and coarse with some fine Fines: yellowish brown	1.9	2.9
Kellaways Sand	Siltstone, friable, bluish green, sandy in parts	0.8+	3.7

GRADING

Mean for deposit *percentages* Fines	Sand	Gravel	Depth below surface (m)	*percentages* Fines $-\frac{1}{16}$	Sand $+\frac{1}{16}-\frac{1}{4}$	$+\frac{1}{4}-1$	$+1-4$	Gravel $+4-16$	$+16-64$	$+64$
6	52	42	1.0–2.0	6	10	26	20	35	3	0
			2.0–2.9	5	11	21	18	34	11	0
			Mean	6	10	23	19	35	7	0

COMPOSITION

Depth below surface (m)	Percentage by weight in gravel (+4–64 mm) fraction: Limestone including chalk	Flint	Ironstone	Sandstone	Quartzite	Others
1.0–2.9	40	17	24	1	17	1

Water level (+7.0 m)+23 ft
Water struck at (+4.4 m)
152 mm percussion
February 1976

Overburden 1.7 m
Mineral 3.7 m
Bedrock 0.9 m+

LOG

Geological classification	Lithology	Thickness m	Depth m
	Soil, brown, sandy	0.4	0.4
River Terrace Deposits (First Terrace)	Clay, firm to indurated, brown to pale brown, sandy in part, with pebbles and carbonised plant remains	1.3	1.7
	Sandy gravel Gravel: fine with some coarse, mainly shelly and oolitic limestone and some ?chalk, with ironstone, angular to subrounded flint and quartzite Sand: medium and coarse with trace fine, quartz and limestone ooliths Fines: orange	3.7	5.4
Kellaways Sand	Sand, finely crystalline limestone in upper 0.1 m, compacted, greenish grey, fine grained with bivalve and belemnite fragments	0.9+	6.3

GRADING

Mean for deposit *percentages* Fines	Sand	Gravel	Depth below surface (m)	*percentages* Fines	Sand			Gravel		
				$-\frac{1}{16}$	$+\frac{1}{16}-\frac{1}{4}$	$+\frac{1}{4}-1$	+1–4	+4–16	+16–64	+64
5	55	40	1.7–2.7	8	14	31	14	27	6	0
			2.7–3.7	5	6	27	26	32	4	0
			3.7–4.7	3	3	27	21	35	11	0
			4.7–5.4	4	4	20	26	35	11	0
			Mean	5	7	27	21	32	8	0

COMPOSITION

Depth below surface (m)	Percentage by weight in gravel (+4–64 mm) fraction: Limestone including chalk	Flint	Ironstone	Sandstone	Quartzite	Others
1.7–5.4	45	14	21	1	15	4

TF 10 NE 27 1502 0661 Brook Farm, Glinton **Sub-block F_1**

Surface level (+5.8 m) +19 ft
Water struck at (+4.2 m)
152 mm percussion
February 1976

Overburden 0.9 m
Mineral 2.7 m
Bedrock 0.9 m+

LOG

Geological classification	Lithology	Thickness m	Depth m
	Soil, yellowish brown, with gravel	0.3	0.3
River Terrace Deposits	Clay, yellow, sandy	0.6	0.9
(First Terrace)	Sandy gravel Gravel: fine with trace coarse, mainly angular to subrounded oolitic and shelly limestone with flint and some ironstone and quartzite Sand: medium and coarse with trace fine Fines: yellow	2.7	3.6
Oxford Clay	Clay, firm, dark grey, fissile with shell fragments	0.9+	4.5

GRADING

Mean for deposit *percentages* Fines	Sand	Gravel	Depth below surface (m)	*percentages* Fines $-\frac{1}{16}$	Sand $+\frac{1}{16}-\frac{1}{4}$	$+\frac{1}{4}-1$	$+1-4$	Gravel $+4-16$	$+16-64$	$+64$
8	52	40	0.9–1.9	8	8	27	19	35	3	0
			1.9–3.0	9	6	27	24	26	8	0
			3.0–3.6	4	4	15	21	49	7	0
			Mean	8	6	24	22	34	6	0

COMPOSITION

Depth below surface (m)	Percentage by weight in gravel (+4–64 mm) fraction: Limestone including chalk	Flint	Ironstone	Sandstone	Quartzite	Others
0.9–3.6	33	31	16	1	17	2

TF 10 NE 28 **1542 0539** **Lincoln Road, Glinton** **Sub-block G_1**

Surface level (+9.1 m) +30 ft
Water not struck
152 mm percussion
February 1976

Overburden 1.6 m
Mineral 0.4 m
Bedrock 1.4 m+

LOG

Geological classification	Lithology	Thickness m	Depth m
	Soil, yellowish brown, with gravel	0.3	0.3
River Terrace Deposits (Second Terrace)	Clay, indurated, yellow with some orange and brown mottling, sandy in parts, with occasional pebbles	1.3	1.6
	'Clayey' sandy gravel Gravel: fine with trace coarse, mainly limestone and flint with some ironstone, quartzite and sandstone Sand: fine to medium, trace coarse Fines: yellow	0.4	2.0
Oxford Clay	Clay, stiff to indurated, dark greyish green mottled brown and yellow, bivalve shells and belemnite fragments throughout	1.0	3.0
?Kellaways Sand	Sandstone, indurated, friable in parts, pale greyish green, silty, with thick-shelled bivalves throughout	0.4+	3.4

GRADING

Mean for deposit *percentages*			Depth below surface (m)	*percentages*						
Fines	Sand	Gravel		Fines	Sand			Gravel		
				$-\frac{1}{16}$	$+\frac{1}{16}-\frac{1}{4}$	$+\frac{1}{4}-1$	+1–4	+4–16	+16–64	+64
19	55	26	1.6–2.0	19	26	22	7	21	5	0

COMPOSITION

Depth below surface (m)	Percentage by weight in gravel (+4–64 mm) fraction					
	Limestone including chalk	Flint	Ironstone	Sandstone	Quartzite	Others
1.6–2.0	45	33	11	1	10	0

TF 10 NE 29 1631 0979 Priory Farm, Deeping St. James **Sub-block A_1**

Surface level (+4.9 m) +16 ft
Water struck at (+3.5 m)
152 mm percussion
February 1976

Overburden 0.7 m
Mineral 3.1 m
Bedrock 1.2 m+

LOG

Geological classification	Lithology	Thickness m	Depth m
	Soil, brown, clayey, with gravel	0.7	0.7
River Terrace Deposits (First Terrace)	Sandy gravel Gravel: fine with trace coarse, mainly rounded limestone with flint, quartzite and ironstone Sand: coarse and medium with trace fine Fines: orange-yellow	3.1	3.8
Oxford Clay	Clay, indurated, dark grey, fissile, with many shell fragments	1.2+	5.0

GRADING

Mean for deposit *percentages*			Depth below surface (m)	*percentages*						
Fines	Sand	Gravel		Fines	Sand			Gravel		
				$-\frac{1}{16}$	$+\frac{1}{16}-\frac{1}{4}$	$+\frac{1}{4}-1$	+1–4	+4–16	+16–64	+64
5	53	42	0.7–1.4	8	6	28	24	32	2	0
			1.4–2.4	5	5	30	22	36	2	0
			2.4–3.4	3	3	22	24	42	6	0
			3.4–3.8	4	5	24	21	39	7	0
			Mean	5	7	23	23	38	4	0

COMPOSITION

Depth below surface (m)	Percentage by weight in gravel (+4–64 mm) fraction					
	Limestone including chalk	Flint	Ironstone	Sandstone	Quartzite	Others
0.7–3.8	55	22	8	0	14	1

TF 10 NE 30 1626 0886 Short Drove, Newborough **Sub-block A_1**

Surface level (+5.8 m) +19 ft
Water struck at (+4.6 m)
152 mm percussion
February 1976

Overburden 1.8 m
Mineral 2.7 m
Bedrock 1.0 m+

LOG

Geological classification	Lithology	Thickness m	Depth m
Alluvium	Clay, indurated, brown mottled with yellow and buff	1.0	1.0
	Silt, soft to very soft, greyish buff becoming black below 1.8 m, sandy	0.8	1.8
River Terrace Deposits (First Terrace)	Gravel Gravel: fine with some coarse, mainly angular to subangular flint with limestone, quartzite and ironstone Sand: coarse and medium with trace fine Fines: buff-grey	2.7	4.5
Oxford Clay	Clay, indurated, dark grey with bivalve shell fragments	1.0+	5.5

GRADING

Mean for deposit *percentages* Fines	Sand	Gravel	Depth below surface (m)	*percentages* Fines $-\frac{1}{16}$	Sand $+\frac{1}{16}-\frac{1}{4}$	$+\frac{1}{4}-1$	$+1-4$	Gravel $+4-16$	$+16-64$	$+64$
5	47	48	1.8–2.8	7	7	24	21	35	6	0
			2.8–3.8	4	4	18	24	39	11	0
			3.8–4.5	5	3	11	23	45	13	0
			Mean	5	5	19	23	39	9	0

COMPOSITION

Depth below surface (m)	Percentage by weight in gravel (+4–64 mm) fraction: Limestone including chalk	Flint	Ironstone	Sandstone	Quartzite	Others
1.8–4.5	27	29	19	1	24	0

TF 10 NE 31 **1645 0787** **Manor Farm, Northborough** **Sub-block F_1**

Surface level (+5.2 m) +17 ft
Water struck at (+3.6 m)
152 mm percussion
February 1976

Overburden 1.0 m
Mineral 4.1 m
Bedrock 0.7 m+

LOG

Geological classification	Lithology	Thickness m	Depth m
	Soil, pale brown, clayey	0.5	0.5
River Terrace Deposits (First Terrace)	Silt, firm, mottled orange-yellow, clayey below 0.7 m, becomes pebbly towards base	0.5	1.0
	Sandy gravel Gravel: fine with some coarse, mainly shelly and oolitic limestone with subangular to subrounded ironstone, flint and quartzite Sand: medium and coarse with trace fine, quartz, chalk and ironstone Fines: yellow to brown	4.1	5.1
Oxford Clay	Clay, stiff to indurated, friable in parts, dark grey to blue, silty with bivalve shell fragments	0.7+	5.8

GRADING

Mean for deposit *percentages*			Depth below surface (m)	*percentages*						
Fines	Sand	Gravel		Fines	Sand			Gravel		
				$-\frac{1}{16}$	$+\frac{1}{16}-\frac{1}{4}$	$+\frac{1}{4}-1$	+1–4	+4–16	+16–64	+64
4	51	45	1.0–1.6	7	11	36	17	26	3	0
			1.6–2.6	7	8	30	21	30	4	0
			2.6–3.6	2	4	24	15	46	9	0
			3.6–4.6	3	2	14	23	39	19	0
			4.6–5.1	5	6	23	24	33	9	0
			Mean	4	6	25	20	36	9	0

COMPOSITION

Depth below surface (m)	Percentage by weight in gravel (+4–64 mm) fraction					
	Limestone including chalk	Flint	Ironstone	Sandstone	Quartzite	Others
1.0–5.1	49	14	24	0	11	2

TF 10 NE 32 1627 0646 Manor House Farm, east of Glinton **Sub-block G_1**

Surface level (+9.4 m) +31 ft
Water not struck
152 mm percussion
February 1976

Overburden 0.8 m
Mineral 1.5 m
Bedrock 0.7 m+

LOG

Geological classification	Lithology	Thickness m	Depth m
	Soil, yellowish brown, clayey with gravel	0.4	0.4
River Terrace Deposits (Second Terrace)	Clay, indurated, orange, sandy	0.4	0.8
	'Clayey' gravel Gravel: fine with some coarse, mainly angular to rounded flint with limestone, ironstone and quartzite Sand: medium and coarse with some fine Fines: orange to yellow	1.5	2.3
Oxford Clay	Clay, stiff, buff-grey becoming blue towards base, with bivalve shell fragments	0.7+	3.0

GRADING

Mean for deposit *percentages*			Depth below surface (m)	*percentages*						
Fines	Sand	Gravel		Fines	Sand			Gravel		
				$-\frac{1}{16}$	$+\frac{1}{16}-\frac{1}{4}$	$+\frac{1}{4}-1$	+1–4	+4–16	+16–64	+64
13	43	44	0.8–1.1	19	19	22	12	18	10	0
			1.1–2.1	10	7	17	16	35	15	0
			2.1–2.3	13	7	19	23	26	12	0
			Mean	13	9	18	16	30	14	0

COMPOSITION

Depth below surface (m)	Percentage by weight in gravel (+4–64 mm) fraction					
	Limestone including chalk	Flint	Ironstone	Sandstone	Quartzite	Others
0.8–2.3	28	40	18	2	12	0

TF 10 NE 33 1630 0540 Fox Covert Road, Glinton **Sub-block G_1**

Surface level (+7.9 m) +26 ft
Water not struck
152 mm percussion
February 1976

Overburden 0.8 m
Mineral 1.1 m
Bedrock 0.8 m+

LOG

Geological classification	Lithology	Thickness m	Depth m
	Soil, clayey, sandy with gravel	0.8	0.8
River Terrace Deposits (Second Terrace)	Sandy gravel; with clay layer between 1.1–1.3 m Gravel: fine with some coarse, mainly angular to rounded flint with quartzite, limestone, ironstone and sandstone Sand: fine and medium with trace coarse Fines: yellow to orange	1.1	1.9
Oxford Clay	Clay, stiff, brownish grey mottled, becoming greyish blue and silty, with shell fragments below 2.0 m	0.8+	2.7

GRADING

Mean for deposit *percentages*			Depth below surface (m)	*percentages*						
Fines	Sand	Gravel		Fines	Sand			Gravel		
				$-\frac{1}{16}$	$+\frac{1}{16}-\frac{1}{4}$	$+\frac{1}{4}-1$	+1–4	+4–16	+16–64	+64
5	56	39	0.8–1.9	5	27	22	7	27	12	0

COMPOSITION

Depth below surface (m)	Percentage by weight in gravel (+4–64 mm) fraction					
	Limestone including chalk	Flint	Ironstone	Sandstone	Quartzite	Others
0.8–1.9	21	41	8	2	28	0

TF 10 NE 34 **1768 0854** **Near Deeping St. James Station** **Sub-block A_1**

Surface level (+3.4 m) +11 ft	Overburden 1.0 m
Water struck at (+1.9 m)	Mineral 2.8 m
152 mm percussion	Bedrock 1.3 m+
February 1976	

Geological classification	Lithology	Thickness m	Depth m
	Soil, brown	0.2	0.2
Alluvium	Clay, indurated, brown mottled orange, pebbly towards base	0.8	1.0
River Terrace Deposits (First Terrace)	Gravel Gravel: fine with some coarse, mainly oolitic and shelly limestone with ironstone, angular to rounded flint and quartzite Sand: medium and coarse with trace fine, quartz with limestone ooliths Fines: orange	2.8	3.8
Oxford Clay	Clay, stiff to indurated, blue becoming grey and fossiliferous below 4.4 m	1.3+	5.1

GRADING

Mean for deposit *percentages* Fines	Sand	Gravel	Depth below surface (m)	*percentages* Fines $-\frac{1}{16}$	Sand $+\frac{1}{16}-\frac{1}{4}$	$+\frac{1}{4}-1$	$+1-4$	Gravel $+4-16$	$+16-64$	$+64$
5	45	50	1.0–1.7	10	9	25	18	35	3	0
			1.7–2.7	5	6	33	15	33	8	0
			2.7–3.7	2	3	13	15	48	19	0
			3.7–3.8	4	6	22	18	34	16	0
			Mean	5	6	23	16	39	11	0

COMPOSITION

Depth below surface (m)	Percentage by weight in gravel (+4–64 mm) fraction: Limestone including chalk	Flint	Ironstone	Sandstone	Quartzite	Others
1.0–3.8	53	14	19	0	14	0

TF 10 NE 35 **1724 0757** **Castle Farm, Northborough** **Sub-block A_1**

Surface level (+4.6 m) +15 ft
Water struck at (+2.9 m)
152 mm percussion
February 1976

Overburden 1.2 m
Mineral 1.3 m
Bedrock 0.9 m+

LOG

Geological classification	Lithology	Thickness m	Depth m
	Soil, brown	0.2	0.2
Alluvium	Clay, firm, brown, silty	0.5	0.7
River Terrace Deposits (First Terrace)	Clay, sandy, mottled buff to orange-red, with carbonised plant remains and angular flint gravel	0.5	1.2
	Sandy gravel Gravel: fine with trace coarse, mainly oolitic limestone with ironstone, quartzite and subangular flint Sand: medium and coarse with some fine Fines: orange to yellow	1.3	2.5
Oxford Clay	Clay, firm to stiff, greenish blue, silty with bivalve shell fragments	0.9+	3.4

GRADING

Mean for deposit *percentages*			Depth below surface (m)	*percentages*						
Fines	Sand	Gravel		Fines	Sand			Gravel		
				$-\frac{1}{16}$	$+\frac{1}{16}-\frac{1}{4}$	$+\frac{1}{4}-1$	+1–4	+4–16	+16–64	+64
8	57	35	1.2–1.7	11	10	25	19	29	6	0
			1.7–2.5	6	6	26	26	30	6	0
			Mean	8	8	26	23	29	6	0

COMPOSITION

Depth below surface (m)	Percentage by weight in gravel (+4–64 mm) fraction					
	Limestone including chalk	Flint	Ironstone	Sandstone	Quartzite	Others
1.2–2.5	56	10	22	0	12	0

TF 10 NE 36 1742 0672 Long Meadow Road, Peakirk **Sub-block A_1**

Surface level (+3.7 m) +12 ft
Water struck at (+2.0 m)
152 mm percussion
January 1976

Overburden 1.4 m
Mineral 1.6 m
Waste 0.6 m
Mineral 0.9 m
Bedrock 0.6 m+

LOG

Geological classification	Lithology	Thickness m	Depth m
Alluvium	Clay, brown, with gravel	0.5	0.5
River Terrace Deposits (First Terrace)	Clay, sandy, mottled orange to greyish brown	0.9	1.4
	a Gravel	1.6	3.0
	Gravel: mainly fine with some coarse, mostly angular to subrounded oolitic and crystalline limestone with sandstone, flint and some ironstone and quartzite		
	Sand: mainly coarse, with medium and trace fine		
	Fines: orange-brown to buff-grey		
	Silt, soft to firm, dark grey	0.6	3.6
	b Gravel as above	0.9	4.5
Oxford Clay	Clay, indurated, mottled grey to brown in upper 0.2 m becoming bluish grey, with abundant fossil remains	0.6+	5.1

GRADING

	Mean for deposit *percentages*			Depth below surface (m)	*percentages*						
	Fines	Sand	Gravel		Fines	Sand			Gravel		
					$-\frac{1}{16}$	$+\frac{1}{16}-\frac{1}{4}$	$+\frac{1}{4}-1$	+1–4	+4–16	+16–64	+64
a	5	40	55	1.4–2.4	5	4	18	24	40	9	0
				2.4–3.0	5	2	11	20	41	21	0
				Mean	5	3	15	22	40	15	0
				3.0–3.6	Waste						
b	5	38	57	3.6–4.5	5	2	14	22	48	9	0
a + b	5	40	55	Mean	5	3	15	22	43	12	0

COMPOSITION **(a + b)**

Depth below surface (m)	Percentage by weight in gravel (+4–64 mm) fraction					
	Limestone including chalk	Flint	Ironstone	Sandstone	Quartzite	Others
1.4–4.5	47	16	7	23	5	2

TF 10 NE 37 1752 0582 Long Meadow Road, Peakirk **Block F**

Surface level (+3.0 m) +10 ft
Water struck at (+1.3 m)
152 mm percussion
January 1976

Waste 1.9 m
Bedrock 1.8 m+

LOG

Geological classification	Lithology	Thickness m	Depth m
	Soil, brown, clayey	0.2	0.2
?Alluvium	Clay, firm, brown to orange, silty, becoming pebbly and sandy below 0.8 m	0.9	1.1
River Terrace Deposits (First Terrace)	'Clayey' sandy gravel Gravel: fine with trace coarse, mainly limestone with subangular to subrounded flint, quartzite and ironstone Sand: medium and coarse with some fine Fines: orange	0.8	1.9
Oxford Clay	Clay, indurated friable in parts, dark brown becoming greyish green below 3.0 m, bivalve shell fragments throughout	1.8+	3.7

GRADING

Mean for deposit *percentages* Fines	Sand	Gravel	Depth below surface (m)	*percentages* Fines $-\frac{1}{16}$	Sand $+\frac{1}{16}-\frac{1}{4}$	$+\frac{1}{4}-1$	$+1-4$	Gravel $+4-16$	$+16-64$	$+64$
11	55	34	1.1–1.7	14	9	29	19	26	3	0
			1.7–1.9	4	5	21	25	42	3	0
			Mean	11	8	26	21	31	3	0

COMPOSITION

Depth below surface (m)	Percentage by weight in gravel (+4–64 mm) fraction: Limestone including chalk	Flint	Ironstone	Sandstone	Quartzite	Others
1.1–1.9	48	27	7	0	15	3

TF 10 NE 38 1819 0946 Cranmore Barn, Deeping St. James **Sub-block F_3**

Surface level (+3.0 m) +10 ft
Water struck at (+1.1 m)
152 mm percussion
February 1976

Overburden 1.6 m
Mineral 1.7 m
Bedrock 1.4 m+

LOG

Geological classification	Lithology	Thickness m	Depth m
	Soil, dark brown, clayey	0.5	0.5
Alluvium	Clay, indurated, dark brown mottled orange-brown with black carbonaceous patches	0.4	0.9
River Terrace Deposits (First Terrace)	Clay, indurated, orange mottled dark brown, sandy, below 1.5 m with gravel of flint and limestone	0.7	1.6
	Sandy gravel Gravel: fine with trace coarse, mainly rounded to subangular limestone, with ironstone, flint and quartzite Sand: coarse, medium with some fine Fines: orange to buff-yellow	1.7	3.3
Oxford Clay	Clay, stiff, greenish blue, silty, trace bivalve shell fragments	1.4+	4.7

GRADING

Mean for deposit *percentages*			Depth below surface (m)	*percentages*						
Fines	Sand	Gravel		Fines	Sand			Gravel		
				$-\frac{1}{16}$	$+\frac{1}{16}-\frac{1}{4}$	$+\frac{1}{4}-1$	+1–4	+4–16	+16–64	+64
5	49	46	1.6–2.0	7	14	27	18	29	5	0
			2.0–3.0	5	7	20	24	39	5	0
			3.0–3.3	2	3	12	20	50	13	0
			Mean	5	8	20	21	39	7	0

COMPOSITION

Depth below surface (m)	Percentage by weight in gravel (+4–64 mm) fraction					
	Limestone including chalk	Flint	Ironstone	Sandstone	Quartzite	Others
1.6–3.3	35	22	29	0	13	1

TF 10 NE 39 1876 0922 Cranmore Lodge, Deeping St. James **Sub-block D_1**

Surface level (+2.4 m) +8 ft
Water struck at (+0.9 m)
152 mm percussion
February 1976

Overburden 1.1 m
Mineral 2.5 m
Bedrock 1.1 m+

LOG

Geological classification	Lithology	Thickness m	Depth m
Made ground		0.4	0.4
River Terrace Deposits (First Terrace)	Clay, indurated, dark brown mottled buff and orange, with black organic patches, pebbly below 0.9 m	0.7	1.1
	Sandy gravel Gravel: fine with trace coarse, mainly rounded to subangular oolitic, shelly and crystalline limestone with quartzite, flint and ironstone Sand: coarse and medium with trace fine, mostly limestone ooliths Fines: grey-buff	2.5	3.6
Oxford Clay	Clay, stiff, blue, silty, with ?silt-infilled burrows	1.1+	4.7

GRADING

Mean for deposit percentages			Depth below surface (m)	percentages						
Fines	Sand	Gravel		Fines	Sand			Gravel		
				$-\frac{1}{16}$	$+\frac{1}{16}-\frac{1}{4}$	$+\frac{1}{4}-1$	+1–4	+4–16	+16–64	+64
7	47	46	1.1–2.1	11	13	24	20	28	4	0
			2.1–3.1	3	3	16	27	46	5	0
			3.1–3.6	6	3	16	13	49	13	0
			Mean	7	7	19	21	40	6	0

COMPOSITION

Depth below surface (m)	Percentage by weight in gravel (+4–64 mm) fraction					
	Limestone including chalk	Flint	Ironstone	Sandstone	Quartzite	Others
1.1–3.6	51	16	14	1	17	1

TF 10 NE 40 1858 0837 Deeping Common, Deeping St. James Sub-block F_3

Surface level (+2.7 m) +9 ft
Water struck at (+0.7 m)
152 mm percussion
February 1976

Overburden 1.2 m
Mineral 3.3 m
Bedrock 1.2 m+

LOG

Geological classification	Lithology	Thickness m	Depth m
River Terrace Deposits (First Terrace)	Clay, stiff, brown mottled with orange, sandy and pebbly below 0.9 m	1.2	1.2
	Gravel Gravel: fine with some coarse, mainly shelly, oolitic and crystalline limestone with rounded to subangular ironstone, flint and quartzite Sand: coarse with medium and fine, mainly limestone ooliths Fines: buff-yellow	3.3	4.5
Oxford Clay	Clay, stiff, blue, silty, trace bivalve shell fragments	1.2+	5.7

GRADING

Mean for deposit *percentages* Fines	Sand	Gravel	Depth below surface (m)	*percentages* Fines	Sand			Gravel		
				$-\frac{1}{16}$	$+\frac{1}{16}-\frac{1}{4}$	$+\frac{1}{4}-1$	+1–4	+4–16	+16–64	+64
4	44	52	1.2–2.1	8	7	34	16	28	7	0
			2.1–3.1	3	2	16	25	45	9	0
			3.1–4.1	3	1	7	32	50	7	0
			4.1–4.5	4	3	10	25	46	12	0
			Mean	4	3	16	25	43	9	0

COMPOSITION

Depth below surface (m)	Percentage by weight in gravel (+4–64 mm) fraction: Limestone including chalk	Flint	Ironstone	Sandstone	Quartzite	Others
1.2–4.5	54	12	25	1	8	0

TF 10 NE 41 **1867 0731** **Moor Drain, Peakirk** **Sub-block A_1**

Surface level (+2.7 m) +9 ft
Water level not recorded
152 mm percussion
January 1976

Overburden 1.4 m
Mineral 0.8 m
Waste 1.4 m
Bedrock 0.4 m+

LOG

Geological classification	Lithology	Thickness m	Depth m
	Soil, sandy and clayey, with gravel	0.4	0.4
Alluvium	Clay, firm, brown, mottled with orange	0.7	1.1
	Clay, firm, orange mottled with yellow, sandy	0.3	1.4
River Terrace Deposits (First Terrace)	'Clayey' sand Gravel: trace fine only Sand: medium and fine with trace coarse Fines: orange-yellow	0.8	2.2
	Silt, with coarse sandy layers and trace flint gravel, firm, blue mottled with pale yellow and buff	1.4	3.6
Oxford Clay	Clay, stiff, bluish grey, silty, some flint pebbles in upper 0.1 m, trace bivalve shell fragments	0.4+	4.0

GRADING

Mean for deposit *percentages*			Depth below surface (m)	*percentages*						
Fines	Sand	Gravel		Fines	Sand			Gravel		
				$-\frac{1}{16}$	$+\frac{1}{16}-\frac{1}{4}$	$+\frac{1}{4}-1$	+1–4	+4–16	+16–64	+64
18	80	2	1.4–2.2	18	32	42	6	2	0	0

TF 10 NE 42 1850 0617 Milkingnook Road, Newborough **Sub-block A_1**

Surface level (+2.7 m) +9 ft
Water level not recorded
152 mm percussion
February 1976

Overburden 1.2 m
Mineral 1.7 m
Bedrock 1.1 m+

LOG

Geological classification	Lithology	Thickness m	Depth m
	Soil, brown, clayey	0.3	0.3
Alluvium	Clay, dark brown to orange, silty with occasional peaty patches, with sand and pebbles below 0.8 m	0.9	1.2
River Terrace Deposits (First Terrace)	'Clayey' sandy gravel Gravel: fine with trace coarse, mainly angular to subrounded flint with limestone and some quartzite, ironstone and sandstone Sand: medium and coarse with trace fine, limestone ooliths and ironstone with ?chalk traces Fines: orange to buff-grey	1.7	2.9
Oxford Clay	Clay, firm to stiff, blue, with silt patches and bivalve shell fragments	1.1+	4.0

GRADING

Mean for deposit *percentages* Fines	Sand	Gravel	Depth below surface (m)	*percentages* Fines $-\frac{1}{16}$	Sand $+\frac{1}{16}-\frac{1}{4}$	$+\frac{1}{4}-1$	$+1-4$	Gravel $+4-16$	$+16-64$	$+64$
11	45	44	1.2–1.8	10	10	34	15	27	4	0
			1.8–2.5	17	6	14	16	39	8	0
			2.5–2.9	4	4	16	17	53	6	0
			Mean	11	7	22	16	38	6	0

COMPOSITION

Depth below surface (m)	Percentage by weight in gravel (+4–64 mm) fraction: Limestone including chalk	Flint	Ironstone	Sandstone	Quartzite	Others
1.2–2.9	33	39	6	6	10	6

TF 10 NE 43 **1856 0542** **Werrington Bridge Road, Newborough** **Block D**

Surface level (+2.1 m)+7 ft
Water struck at (−0.1 m)
152 mm percussion
January 1976

Waste 2.8 m
Bedrock 1.0 m+

LOG

Geological classification	Lithology	Thickness m	Depth m
	Soil, dark brown, peaty	0.6	0.6
Nordelph Peat	Peat, dark brown, silty	0.8	1.4
Barroway Drove Beds	Silt, soft, dark bluish grey with patches of peat	0.8	2.2
River Terrace Deposits (First Terrace)	Sandy gravel Gravel: mainly fine with some coarse, angular to subrounded limestone, ironstone, flint and sandstone Sand: mainly coarse and medium with trace fine Fines: greyish buff	0.6	2.8
Oxford Clay	Clay, stiff, greenish grey, silty with bivalve shell fragments	1.0+	3.8

GRADING

Mean for deposit *percentages*			Depth below surface (m)	*percentages*						
Fines	Sand	Gravel		Fines	Sand			Gravel		
				$-\frac{1}{16}$	$+\frac{1}{16}-\frac{1}{4}$	$+\frac{1}{4}-1$	+1–4	+4–16	+16–64	+64
6	35	59	2.2–2.8	6	2	13	20	47	12	0

TF 10 NE 44 **1966 0814** **Decoy Farm, Newborough** **Sub-block A_1**

Surface level (+3.0 m) +10 ft
Water struck at (+0.5 m)
152 mm percussion
January 1976

Waste 2.7 m
Bedrock 1.5 m+

LOG

Geological classification	Lithology	Thickness m	Depth m
Alluvium	Soil, brown, mottled orange, clayey	0.9	0.9
?Nordelph Peat	Clay, dark brown to black, peaty, silty in places	0.5	1.4
Barroway Drove Beds	Silt, dark grey to blue, with peat pockets	0.3	1.7
River Terrace Deposits (First Terrace)	Clay, grey mottled yellow to white, occasional carbonised plant remains, becomes sandy below 2.5 m	1.0	2.7
Oxford Clay	Clay, firm to stiff, grey mottled yellow, selenite crystals, fragments of ammonites and bivalve shells common	1.5+	4.2

TF 10 NE 45 1963 0722 Moor Farm, Newborough **Sub-block F_3**

Surface level (+3.3 m) +11 ft
Water struck at (+1.5 m)
152 mm percussion
January 1976

Overburden 1.1 m
Mineral 1.6 m
Bedrock 1.4 m+

LOG

Geological classification	Lithology	Thickness m	Depth m
	Soil, brown, clayey	0.4	0.4
River Terrace Deposits (First Terrace)	Clay, firm, brown to orange, silty in parts, trace pebbles especially below 0.9 m	0.7	1.1
	'Very clayey' pebbly sand Gravel: trace amounts of fine and coarse, flint Sand: medium with fine and trace coarse Fines: orange to yellow	1.6	2.7
Oxford Clay	Clay, stiff to firm, grey with yellow mottling, trace shell fragments, selenite pockets and ?silt infilled burrows; becomes blue and unfossiferous below 3.4 m	1.4+	4.1

GRADING

Mean for deposit *percentages*			Depth below surface (m)	*percentages*						
Fines	Sand	Gravel		Fines	Sand			Gravel		
				$-\frac{1}{16}$	$+\frac{1}{16}-\frac{1}{4}$	$+\frac{1}{4}-1$	+1–4	+4–16	+16–64	+64
22	72	6	1.1–1.8	35	19	26	10	10	0	0
			1.8–2.7	12	20	60	4	2	2	0
			Mean	22	20	45	7	5	1	0

TF 10 NE 46 1953 0635 Slip Bridge, Newborough **Block D**

Surface level (+1.2 m) +4 ft
Water struck at (−0.6 m)
152 mm percussion
February 1976

Waste 2.6 m
Bedrock 1.0 m+

LOG

Geological classification	Lithology	Thickness m	Depth m
	Soil, brown, clayey	0.2	0.2
Alluvium	Clay, firm, brown, slightly mottled orange	0.4	0.6
Nordelph Peat	Peat, firm, dark brown, clayey, silty in places	0.7	1.3
Barroway Drove Beds	Silt, soft, blue mottled buff, yellow and dark grey	0.5	1.8
River Terrace Deposits (First Terrace)	'Clayey' pebbly sand Gravel: fine with trace coarse, mainly angular to rounded limestone with sandstone, flint, quartzite and ironstone Sand: medium with coarse and trace fine Fines: yellow	0.8	2.6
Oxford Clay	Clay, firm to stiff, bluish grey, silty with trace bivalve shell fragments	1.0+	3.6

GRADING

Mean for deposit *percentages*			Depth below surface (m)	*percentages*						
Fines	Sand	Gravel		Fines	Sand			Gravel		
				$-\frac{1}{16}$	$+\frac{1}{16}-\frac{1}{4}$	$+\frac{1}{4}-1$	+1–4	+4–16	+16–64	+64
10	70	20	1.8–2.6	10	7	45	18	16	4	0

COMPOSITION

Depth below surface (m)	Percentage by weight in gravel (+4–64 mm) fraction					
	Limestone including chalk	Flint	Ironstone	Sandstone	Quartzite	Others
1.8–2.6	46	20	3	24	7	0

TF 10 NE 47 1966 0544 Middle Road, Newborough

Surface level (+1.2 m) +4 ft
Water level not recorded
152 mm percussion
February 1976

Waste 2.0 m
Bedrock 1.8 m+

LOG

Geological classification	Lithology	Thickness m	Depth m
Nordelph Peat	Soil, dark brown, peaty	0.6	0.6
Barroway Drove Beds	Clay, brown, mottled with dark brown and orange	0.3	0.9
	Silt, soft, blue mottled pale grey, sandy in part with some pebbles at base		
River Terrace Deposits (First Terrace)	Sandy gravel Gravel: fine with trace coarse, limestone and flint with quartzite and ironstone Sand: medium and coarse with trace fine Fines: buff-yellow	0.6	2.0
Oxford Clay	Clay, stiff, friable in places, bluish grey, traces of fossils throughout	1.8+	3.8

GRADING

Mean for deposit *percentages* Fines	Sand	Gravel	Depth below surface (m)	*percentages* Fines	Sand			Gravel		
				$-\frac{1}{16}$	$+\frac{1}{16}-\frac{1}{4}$	$+\frac{1}{4}-1$	+1–4	+4–16	+16–64	+64
9	68	23	1.4–2.0	9	8	36	24	22	1	0

COMPOSITION

Depth below surface (m)	Percentage by weight in gravel (+4–64 mm) fraction Limestone including chalk	Flint	Ironstone	Sandstone	Quartzite	Others
1.4–2.0	39	39	6	1	15	0

TF 10 NE 48 **1718 0955** **Priory Farm, Deeping St. James** **Sub-block A_1**

Surface level (+4.0 m) +13 ft
Water struck at (+1.9 m)
152 mm percussion
February 1976

Overburden 1.5 m
Mineral 2.2 m
Bedrock 0.8 m+

LOG

Geological classification	Lithology	Thickness m	Depth m
	Soil, brown, clayey with pebbles	0.4	0.4
Alluvium	Clay, stiff, brown mottled with dark grey and orange; becomes indurated, pale yellow to orange, sandy and pebbly below 0.9 m	1.1	1.5
River Terrace Deposits (First Terrace)	Gravel Gravel: fine with some coarse, rounded limestone, subangular flint, ironstone and quartzite with some sandstone Sand: medium and coarse with trace fine, quartz and limestone ooliths Fines: orange to buff-grey	2.2	3.7
Oxford Clay	Clay, stiff to indurated, blue, silty in part, with bivalve and ammonite fragments	0.8+	4.5

GRADING

Mean for deposit *percentages*			Depth below surface (m)	*percentages*						
Fines	Sand	Gravel		Fines	Sand			Gravel		
				$-\frac{1}{16}$	$+\frac{1}{16}-\frac{1}{4}$	$+\frac{1}{4}-1$	+1–4	+4–16	+16–64	+64
5	46	49	1.5–2.5	10	8	20	19	34	9	0
			2.5–3.5	3	3	23	25	39	7	0
			3.5–3.7	4	3	20	17	49	7	0
			Mean	5	5	21	20	41	8	0

COMPOSITION

Depth below surface (m)	Percentage by weight in gravel (+4–64 mm) fraction					
	Limestone including chalk	Flint	Ironstone	Sandstone	Quartzite	Others
1.5–3.7	34	28	16	4	18	0

TF 10 NE 49 1771 0537 Long Meadow Road, Peakirk **Block F**

Surface level unknown
Water struck at 1.7 m below surface
76 mm Minuteman
August 1977

Waste 2.0 m
Bedrock 3.0 m +

LOG

Geological classification	Lithology	Thickness m	Depth m
	Soil, dark greyish brown	0.3	0.3
River Terrace Deposits (First Terrace)	Clay, firm, variegated pale brown with pale orange-brown and grey, calcareous, sandy	0.3	0.6
	Clay, khaki-brown to grey with trace angular to subangular flint, limestone and sandstone gravel with coarse to fine sand	1.4	2.0
Oxford Clay	Clay, weathered between 2.0–4.0 m, firm to stiff, pale to dark grey, micaceous, calcareous and trace shell fragments	3.0 +	5.0

TF 10 SW 10 1390 0481 Near Woodcroft Castle, Etton **Sub-block F_1**

Surface level (+7.3 m) +24 ft
Water level not recorded
152 mm percussion
February 1976

Overburden 1.7 m
Mineral 0.5 m
Bedrock 1.4 m +

LOG

Geological classification	Lithology	Thickness m	Depth m
	Soil, brown, clayey	0.3	0.3
River Terrace Deposits (First Terrace)	Clay, stiff, pale brown, some flint gravel	1.4	1.7
	'Very clayey' pebbly sand Gravel: fine, mainly flint and limestone Sand: fine to medium, some coarse, mainly quartz Fines: pale orange-brown	0.5	2.2
Blisworth Clay	Clay, stiff, dark grey, silty with ironstaining in upper 0.9 m, interbedded with indurated finely crystalline limestone below 3.4 m	1.4 +	3.6

GRADING

Mean for deposit *percentages*			Depth below surface (m)	*percentages*						
Fines	Sand	Gravel		Fines	Sand			Gravel		
				$-\frac{1}{16}$	$+\frac{1}{16}-\frac{1}{4}$	$+\frac{1}{4}-1$	+1–4	+4–16	+16–64	+64
37	55	8	1.7–2.2	37	27	21	7	8	0	0

TF 10 SW 11 **1371 0140** **Belsize Farm, Marholm** **Block O**

Surface level (+29.5 m) +97 ft
Water struck at (+23.5 m)
152 mm percussion
February 1976

Overburden 0.5 m
Mineral 1.9 m
Bedrock 4.5 m+

LOG

Geological classification	Lithology	Thickness m	Depth m
	Soil, brown	0.2	0.2
Glacial Sand and Gravel	Clay, orange-brown, sandy with flint pebbles	0.3	0.5
	'Clayey' sandy gravel Gravel: fine and coarse with trace cobble, mainly angular to subangular flint and limestone, with subrounded sandstone, some ironstone and quartzite Sand: medium with some coarse and fine, quartz and flint Fines: orange-brown	1.9	2.4
?Cornbrash	Clay, firm to stiff, dark grey to bluish green, calcareous with trace shell fragments, passes into well indurated limestone below 6.0 m	4.5+	6.9

GRADING

Mean for deposit *percentages*			Depth below surface (m)	*percentages*						
Fines	Sand	Gravel		Fines	Sand			Gravel		
				$-\frac{1}{16}$	$+\frac{1}{16}-\frac{1}{4}$	$+\frac{1}{4}-1$	+1–4	+4–16	+16–64	+64
14	43	43	0.5–2.4	14	10	22	11	21	18	4

COMPOSITION

Depth below surface (m)	Percentage by weight in gravel (+4–64 mm) fraction					
	Limestone including chalk	Flint	Ironstone	Sandstone	Quartzite	Others
0.5–2.4	36	47	2	10	2	3

TF 10 SW 12 **1497 0329** **Near Middle Farm, Marholm** **Sub-block G_1**

Surface level (+8.8 m) +29 ft
Water level not recorded
152 mm percussion
February 1976

Overburden 1.0 m
Mineral 1.6 m
Bedrock 0.2 m+

LOG

Geological classification	Lithology	Thickness m	Depth m
	Soil	0.3	0.3
River Terrace Deposits (Second Terrace)	Clay, firm to stiff, pale brown, with flint gravel	0.7	1.0
	'Clayey' sandy gravel Gravel: fine with some coarse and trace cobble, mainly limestone with flint, some ironstone and quartzite, with occasional shell fragments Sand: fine to coarse, mainly quartz Fines: pale brown to dark brown	1.6	2.6
?Kellaways Sand	Clay, firm to very well indurated, dark greyish blue, silty, with many shell fragments	0.2+	2.8

GRADING

Mean for deposit *percentages*			Depth below surface (m)	*percentages*						
Fines	Sand	Gravel		Fines	Sand			Gravel		
				$-\frac{1}{16}$	$+\frac{1}{16}-\frac{1}{4}$	$+\frac{1}{4}-1$	+1–4	+4–16	+16–64	+64
10	47	43	1.0–2.6	10	13	17	17	35	7	1

COMPOSITION

Depth below surface (m)	Percentage by weight in gravel (+4–64 mm) fraction					
	Limestone including chalk	Flint	Ironstone	Sandstone	Quartzite	Others
1.0–2.6	77	17	2	1	2	1

TF 10 SE 785 **1574 0432** **Canon's Barn, Werrington, Peterborough** **Sub-block G_1**

Surface level (+11.3 m) +37 ft
Water level not recorded
152 mm percussion
February 1976

Overburden 0.6 m
Mineral 2.4 m
Bedrock 0.6 m+

LOG

Geological classification	Lithology	Thickness m	Depth m
Made ground	Mixture of bricks, gravel and ash	0.2	0.2
River Terrace Deposits (Second Terrace)	Clay, brown, sandy towards base with flint gravel	0.4	0.6
	'Very clayey' pebbly sand Gravel: trace amounts of fine and coarse, subrounded sandstone and subangular flint Sand: fine with medium and trace coarse, quartz and flint Fines: pale orange-brown	2.4	3.0
Oxford Clay	Clay, stiff, dark grey, silty, trace shell fragments	0.6+	3.6

GRADING

Mean for deposit *percentages*			Depth below surface (m)	*percentages*						
Fines	Sand	Gravel		Fines	Sand			Gravel		
				$-\frac{1}{16}$	$+\frac{1}{16}-\frac{1}{4}$	$+\frac{1}{4}-1$	+1–4	+4–16	+16–64	+64
39	56	5	0.6–3.0	39	30	23	3	3	2	0

TF 10 SE 786 1522 0290 North of Marholm, Peterborough **Sub-block G_1**

Surface level (+10.0 m)+33 ft
Water not struck
152 mm percussion
February 1976

Overburden 0.4 m
Mineral 0.8 m
Bedrock 1.4 m+

LOG

Geological classification	Lithology	Thickness m	Depth m
	Soil, brown, sandy with some gravel	0.4	0.4
River Terrace Deposits (Second Terrace)	'Clayey' pebbly sand Gravel: fine with trace coarse, predominantly limestone with trace amounts of flint, quartzite, ironstone and shell fragments Sand: medium with some fine and trace coarse Fines: pale orange-brown	0.8	1.2
Oxford Clay	Clay, soft to firm becoming stiff, dark khaki-grey, traces of shell fragments	1.4+	2.6

GRADING

Mean for deposit *percentages*			Depth below surface (m)	*percentages*						
Fines	Sand	Gravel		Fines	Sand			Gravel		
				$-\frac{1}{16}$	$+\frac{1}{16}-\frac{1}{4}$	$+\frac{1}{4}-1$	+1–4	+4–16	+16–64	+64
13	72	15	0.4–1.2	13	16	51	5	12	3	0

COMPOSITION

Depth below surface (m)	Percentage by weight in gravel (+4–64 mm) fraction					
	Limestone including chalk	Flint	Ironstone	Sandstone	Quartzite	Others
0.4–1.2	81	7	2	0	5	5

TF 10 SE 788 1779 0463 Werrington Bridge Road, Peterborough **Block F**

Surface level (+4.5 m)+15 ft
Water not struck
152 mm percussion
February 1976

Waste 2.1 m
Bedrock 0.9 m+

LOG

Geological classification	Lithology	Thickness m	Depth m
	Soil, brown with flint gravel	0.4	0.4
River Terrace Deposits (First Terrace)	Clay, firm, orange-brown mottled pale brown, sandy in part with flint and ironstone pebbles	0.7	2.1
Oxford Clay	Clay, stiff, silty, greyish brown with shell fragments	0.9+	3.0

TF 10 SE 790 **1823 0422** **Werrington End Farm, Werrington, Peterborough** **Block F**

Surface level (+2.1 m)+7 ft
Water not struck
152 mm percussion
February 1976

Waste 2.1 m
Bedrock 1.9 m+

LOG

Geological classification	Lithology	Thickness m	Depth m
	Soil, dark brown, peaty with white shell fragments, ironstained and clayey at base	0.7	0.7
River Terrace Deposits (First Terrace)	**a** 'Clayey' gravel Gravel: fine with some coarse and fine, mainly subangular to subrounded flint with sandstone, trace shell fragments Sand: medium and coarse with some fine, mainly quartz Fines: orange-brown	0.5	1.2
	Clay, very stiff, mottled grey to orange-brown, silty, some pebbles	0.4	1.6
	b 'Clayey' gravel—as above	0.5	2.1
Oxford Clay	Clay, stiff, grey to fawn with many shell fragments, passes into a thin black vitreous, carbonaceous layer at about 3.0 m; becoming dark grey to bluish grey below 3.8 m	1.9+	4.0

GRADING

	Mean for deposit *percentages*			Depth below surface (m)	*percentages*						
	Fines	Sand	Gravel		Fines	Sand			Gravel		
					$-\frac{1}{16}$	$+\frac{1}{16}-\frac{1}{4}$	$+\frac{1}{4}-1$	+1–4	+4–16	+16–64	+64
a	15	37	48	0.7–1.2	15	10	15	12	28	12	8
				1.2–1.6	waste						
b	19	41	40	1.6–2.1	19	6	20	15	27	9	4
a+b	17	39	44	Mean	17	8	18	13	27	9	8

TF 10 SE 791 1884 0363 Near Gunthorpe, Peterborough **Block D**

Surface level (+3.6 m) +12 ft
Water not struck
152 mm percussion
February 1976

Waste 1.6 m
Bedrock 1.4 m+

LOG

Geological classification	Lithology	Thickness m	Depth m
Made ground	Mixture of clay, peat, brick fragments and gravel	0.5	0.5
River Terrace Deposits (?First Terrace)	Clay, greyish brown to yellowish brown, sandy in part, trace gravel, thin sandy layer at base	1.1	1.6
Oxford Clay	Clay, mottled grey to brown, pockets of corroded selenite crystals, shell fragments common; becomes bluish grey below 2.9 m	1.4+	3.0

TF 10 SE 792 1963 0452 The Firs, Newborough Fen, Peterborough **Block D**

Surface level (+1.2 m) +4 ft
Water not struck
152 mm percussion
February 1976

Waste 2.9 m
Bedrock 1.5 m+

LOG

Geological classification	Lithology	Thickness m	Depth m
	Soil, dark brown to black, peaty	0.4	0.4
Nordelph Peat	Peat, friable, black	0.5	0.9
Barroway Drove Beds	Silt, soft, glutinous, pale grey to fawn, with peat fragments	1.7	2.6
River Terrace Deposits (First Terrace)	Clay, sandy, orange-brown mottled pale grey, trace gravel	0.3	2.9
Oxford Clay	Clay, stiff, brown to greyish brown, with abundant shell fragments below 4.1 m	1.5+	4.4

TF 10 SE 793 1942 0357 Near Gunthorpe Bridge, Peterborough **Block F**

Surface level (+3.0 m) +10 ft
Water not struck
152 mm percussion
February 1976

Waste 1.4 m
Bedrock 2.1 m+

LOG

Geological classification	Lithology	Thickness m	Depth m
	Soil, greyish brown	0.5	0.5
River Terrace Deposits (First Terrace)	Silt, soft, variegated pale grey to fawnish brown, silty, trace root fragments, occasional dark brown to black pockets, fine quartz sand layer below 1.3 m	0.9	1.4
Oxford Clay	Clay, stiff, pale grey to khaki becoming bluish grey, mottled orange-brown in patches, with corroded selenite crystals throughout, some shell fragments	2.1+	3.5

TF 10 SE 818 1832 0478 Bridgehill Road, Newborough Fen, Peterborough **Block F**

Surface level (c + 5 m) c + 15 ft
Water struck at c O.D.
76 mm Minuteman
August 1977

Waste 3.0 m
Bedrock 3.0 m +

LOG

Geological classification	Lithology	Thickness m	Depth m
	Soil, brown, clayey below 0.2 m with trace sand and gravel	0.8	0.8
?River Terrace Deposits (First Terrace)	Clay, soft to firm, variegated pale bluish grey with pale orange, occasional small white calcareous pockets of ?corroded selenite crystals	2.2	3.0
Oxford Clay	Clay, firm, dark grey, trace shell fragments	3.0 +	6.0

APPENDIX G

LIST OF SAND AND GRAVEL PITS

Present operator (1977)	*Location*	*Geological formation*	*Grid reference*
ACTIVE			
Redland Aggregates Ltd. (Gravel)	Bainton	Alluvium overlying First Terrace and exposed First Terrace	109 065
Redland Aggregates Ltd. (Gravel Readymix)	West Deeping	Exposed First Terrace	105 095
Hoveringham Gravels Ltd.	Maxey	Exposed First Tarrace	125 078
Nene Barge and Lighter Co. Ltd	Wansford	Alluvium overlying First Terrace and exposed First Terrace	050 072
DISUSED			
Unknown	Northborough	Exposed First Terrace	180 080

APPENDIX H

CONVERSION TABLE, METRES TO FEET (TO NEAREST 0.5 FT)

m	ft	m	ft	m	ft	m	ft	m	ft
0.1	0.5	6.1	20	12.1	39.5	18.1	59.5	24.1	79
0.2	0.5	6.2	20.5	12.2	40	18.2	59.5	24.2	79.5
0.3	1	6.3	20.5	12.3	40.5	18.3	60	24.3	79.5
0.4	1.5	6.4	21	12.4	40.5	18.4	60.5	24.4	80
0.5	1.5	6.5	21.5	12.5	41	18.5	60.5	24.5	80.5
0.6	2	6.6	21.5	12.6	41.5	18.6	61	24.6	80.5
0.7	2.5	6.7	22	12.7	41.5	18.7	61.5	24.7	81
0.8	2.5	6.8	22.5	12.8	42	18.8	61.5	24.8	81.5
0.9	3	6.9	22.5	12.9	42.5	18.9	62	24.9	81.5
1.0	3.5	7.0	23	13.0	42.5	19.0	62.5	25.0	82
1.1	3.5	7.1	23.5	13.1	43	19.1	62.5	25.1	82.5
1.2	4	7.2	23.5	13.2	43.5	19.2	63	25.2	82.5
1.3	4.5	7.3	24	13.3	43.5	19.3	63.5	25.3	83
1.4	4.5	7.4	24.5	13.4	44	19.4	63.5	25.4	83.5
1.5	5	7.5	24.5	13.5	44.5	19.5	64	25.5	83.5
1.6	5	7.6	25	13.6	44.5	19.6	64.5	25.6	84
1.7	5.5	7.7	25.5	13.7	45	19.7	64.5	25.7	84.5
1.8	6	7.8	25.5	13.8	45.5	19.8	65	25.8	84.5
1.9	6	7.9	26	13.9	45.5	19.9	65.5	25.9	85
2.0	6.5	8.0	26	14.0	46	30.0	65.5	26.0	85.5
2.1	7	8.1	26.5	14.1	46.5	20.1	66	26.1	85.5
2.2	7	8.2	27	14.2	46.5	20.2	66.5	26.2	86
2.3	7.5	8.3	27	14.3	47	20.3	66.5	26.3	86.5
2.4	8	8.4	27.5	14.4	47	20.4	67	26.4	86.5
2.5	8	8.5	28	14.5	47.5	20.5	67.5	26.5	87
2.6	8.5	8.6	28	14.6	48	20.6	67.5	26.6	87.5
2.7	9	8.7	28.5	14.7	48	20.7	68	26.7	87.5
2.8	9	8.8	29	14.8	48.5	20.8	68	26.8	88
2.9	9.5	8.9	29	14.9	49	20.9	68.5	26.9	88.5
3.0	10	9.0	29.5	15.0	49	21.0	69	27.0	88.5
3.1	10	9.1	30	15.1	49.5	21.1	69	27.1	89
3.2	10.5	9.2	30	15.2	50	21.2	69.5	27.2	89
3.3	11	9.3	30.5	15.3	50	21.3	70	27.3	89.5
3.4	11	9.4	31	15.4	50.5	21.4	70	27.4	90
3.5	11.5	9.5	31	15.5	51	21.5	70.5	27.5	90
3.6	12	9.6	31.5	15.6	51	21.6	71	27.6	90.5
3.7	12	9.7	32	15.7	51.5	21.7	71	27.7	91
3.8	12.5	9.8	32	15.8	52	21.8	71.5	27.8	91
3.9	13	9.9	32.5	15.9	52	21.9	72	27.9	91.5
4.0	13	10.0	33	16.0	52.5	22.0	72	28.0	92
4.1	13.5	10.1	33	16.1	53	22.1	72.5	28.1	92
4.2	14	10.2	33.5	16.2	53	22.2	73	28.2	92.5
4.3	14	10.3	34	16.3	53.5	22.3	73	28.3	93
4.4	14.5	10.4	34	16.4	54	22.4	73.5	28.4	93
4.5	15	10.5	34.5	16.5	54	22.5	74	28.5	93.5
4.6	15	10.6	35	16.6	45.5	22.6	74	28.6	94
4.7	15.5	10.7	35	16.7	55	22.7	74.5	28.7	94
4.8	15.5	10.8	35.5	16.8	55	22.8	75	28.8	94.5
4.9	16	10.9	36	16.9	55.5	22.9	75	28.9	95
5.0	16.5	11.0	36	17.0	56	23.0	75.5	29.0	95
5.1	17	11.1	36.5	17.1	56	23.1	76	29.1	95.5
5.2	17	11.2	36.5	17.2	56.5	23.2	76	29.2	96
5.3	17.5	11.3	37	17.3	57	23.3	76.5	29.3	96
5.4	17.5	11.4	37.5	17.4	57	23.4	77	29.4	96.5
5.5	18	11.5	37.5	17.5	57.5	23.5	77	29.5	97
5.6	18.5	11.6	38	17.6	57.5	23.6	77.5	29.6	97
5.7	18.5	11.7	38.5	17.7	58	23.7	78	29.7	97.5
5.8	19	11.8	38.5	17.8	58.5	23.8	78	29.8	98
5.9	19.5	11.9	39	17.9	58.5	23.9	78.5	29.9	98
6.0	19.5	12.0	39.5	18.0	59	24.0	78.5	30.0	98.5

REFERENCES

Allen, V. T. 1936. Terminology of medium-grained sediments. *Rep. Natl. Res. Counc. Washington, 1935–1936, App. 1, Rep. Comm. Sedimen.*, 18–47.

Anon. 1970. *Peterborough Development Corporation structure map.* (Peterborough: Development Corporation.)

Archer, A.A. 1969. Background and problems of an assessment of sand and gravel resources in the United Kingdom. *Proc. 9th Commonw. Min. Metall. Congr., 1969*, Vol. 2, Mining and petroleum geology, 495–508. (London: Institution of Mining and Metallurgy.)

— 1970a. Standardisation of the size classification of naturally occurring particles. *Géotechnique*, Vol. 20, 103–107.

— 1970b Making the most of metrication. *Quarry Managers' J.*, Vol. 54, No. 6, 223–227.

Astbury, A. K. 1958 *The Black Fens.* (Cambridge: Golden Head Press.)

Atterberg, A. 1905. Die rationelle Klassifikation der Sande und Kiese. *Chem. Ztg.*, Vol. 29, 195–198.

Biometrika. 1962 *Tables for statisticians*, Vol. 1. 2nd Edition. (Cambridge University Press.)

British Standard 1377. 1967. *Methods of testing soils for civil engineering purposes.* (London: British Standard Instutution.)

Bureau of Mines and Geological Survey. 1948. *Mineral Resources of the United States*, 14–17. (Washington DC: Public Affairs Press.)

Darby, H. C. 1940. *The Draining of the Fens.* (Cambridge University Press.)

Figg, J. 1977. Flint: quality decrease brings need for more caution. *New Civil Eng.* Issue 269, 58–60.

Fillenham, L. F. 1963. Holme Fen Post. *Geogr. J.*, Vol. 129, 502–503.

Fowler, G. 1933. Shrinkage of the peat-covered Fenlands. *Geogr. J.*, Vol. 81, 149–150.

Gallois, R. W. 1979. Geological investigations for the Wash Water Storage Scheme. *Rep. Inst. Geol. Sci.*, No. 78/19.

Godwin, H. 1978. *Fenland: its ancient past and uncertain future.* (Cambridge University Press.)

Godwin, H. and Clifford, H. H. 1938–1940. Studies of the post-glacial history of British Vegetation, Pts 1 to 4. *Phil. Trans. R. Soc. London*, Ser. B.

Hains, B. A. and Horton, A. 1969. *British Regional Geology, Central England*, 3rd Edition. (London: HMSO for Institute of Geological Sciences.)

Hancock, T. and Hawkes, J. 1967. *Greater Peterborough draft basic plan.* (Peterborough: Development Corporation.)

Harris, P. M., Thurrell, R. G., Healing, R. A. and Archer, A.A. 1974. Aggregates in Britian. *Proc. R. Soc. London*, Ser. A, Vol. 339, 329–353.

Hollingworth, S. E., Taylor, J. H. and Kellaway, G. A. 1944. Large scale superficial structures in the Northampton Ironstone Field. *Q. J. Geol. Soc. London*, Vol. 100, 1–44.

Horswill, P. and Horton, A. 1976. Cambering and valley bulging in the Gwash valley at Empingham, Rutland. *Phil. Trans. R. Soc. London*, Ser. A, Vol. 283, 427–462.

Horton, A., Lake, R. D., Bisson, G. and Coppack, B. C. 1974. The geology of Peterborough. *Rep. Inst. Geol. Sci.*, No. 73/12. 86 pp.

Lane, E. W. and others. 1947. Report of the sub-committee on sediment terminology. *Trans. Am. Geophys. Union*, Vol. 28, 936–938.

Pettijohn, F. J. 1957. *Sedimentary rocks.* 2nd Edition. (London: Harper and Row.)

Richardson, S. J. and Smith, J. 1977. Peat wastage in the East Anglian Fens. *J. Soil Sci.*, Vol. 28, 485–589.

Robinson, D. N. 1968. Soil erosion by wind in Lincolnshire, March 1968. *East Midlands Geogr.*, Vol. 4, 351–362.

Roeder, A. R. 1977. Some properties of flint particles and their behaviour in concrete. *Mag. Concr. Res. (Slough)*, Vol. 29, 92–99.

Sabine, P. A. 1949. The source of some erratics from north eastern Northamptonshire and adjacent parts of Huntingdonshire. *Geol. Mag.*, Vol. 86, No. 4, 255–260.

Seale, R. S. 1975. Soils of the Chatteris district of Cambridgeshire [Sheet TL 83]. *Soil Surv. Spec. Surv.*, No. 9.

Skertchly, S. B. J. 1877. The geology of the Fenland. *Mem. Geol. Surv. G.B.*

Taylor, J. H. 1947. Evidence of submarine erosion in the Lincolnshire Limestone of Northamptonshire. *Proc. Geol. Assoc.*, Vol. 57. 246–262.

— 1949. Petrology of the Northampton Sand Ironstone Formation. *Mem. Geol. Surv. G.B.*

Thurrell, R. G. 1971. The assessment of mineral resources with particular reference to sand and gravel. *Quarry Managers' J.*, Vol. 55, pp. 19–25.

Twenhofel, W. H. 1937. Terminology of the fine-grained mechanical sediments. *Rep. Natl. Res. Counc. Washington, 1936–1937, App. 1, Rep. Comm. Sediment.*, 81–104.

Udden, J. A. 1914. Mechanical composition of clastic sediments. *Bull. Geol. Soc. Am.*, Vol. 25, 655–744.

Wentworth, C. K. 1922. A scale of grade and class terms for clastic sediments. *J. Geol.* Vol. 30, No. 5, 377–392.

— 1935. The terminology of coarse sediments. *Bull. Natl. Res. Counc. Washington*, No. 98, 225–246.

Willis, E. H. 1961. Marine transgression sequences in the English Fenlands. *Ann. New York Acad. Sci.*, Vol. 95. 368–376.

Willman, H. B. 1942. Geology and mineral resources of Marseilles, Ottawa and Streator quadrangles. *Bull. Illinois State Geol. Surv.*, No. 66, 343–344.

The following reports of the Institute relate particularly to bulk mineral resources

Reports of the Institute of Geological Sciences

Assessment of British Sand and Gravel Resources

1 The sand and gravel resources of the country south-east of Norwich, Norfolk: Resource sheet TG 20. E.F.P. Nickless.
Report 71/20 ISBN 0 11 880216 £1.15

2 The sand and gravel resources of the country around Witham, Essex: Resource sheet TL 81. H.J.E. Haggard.
Report 71/6 ISBN 0 11 880588 6 £1.20

3 The sand and gravel resources of the area south and west of Woodbridge, Suffolk: Resource sheet TM 24. R. Allender and S. E. Hollyer.
Report 72/9 ISBN 0 11 880596 7 £1.70

4. The sand and gravel resources of the country around Maldon, Essex: Resource sheet TL 80. J. D. Ambrose.
Report 73/1 ISBN 0 11 880600 9 £1.20

5. The sand and gravel resources of the country around Hethersett, Norfolk: Resource sheet TG 10. E. F. P. Nickless.
Report 73/4 ISBN 0 11 880606 8 £1.60

6 The sand and gravel resources of the country around Terling, Essex: Resource sheet TL 71. C. H. Eaton.
Report 73/5 ISBN 0 11 880608 4 £1.20

7. The sand and gravel resources of the country around Layer Breton and Tolleshunt D'Arcy, Essex: Resource sheet TL 91 and part of TL 90. J. D. Ambrose.
Report 73/8 ISBN 0 11 880614 9 £1.30

8 The sand and gravel resources of the country around Shotley and Felixstowe, Suffolk: Resource sheet TM 23. R. Allender and S. E. Hollyer.
Report 73/13 ISBN 0 11 880625 4 £1.60

9 The sand and gravel resources of the country around Attlebridge, Norfolk: Resource sheet TG 11. E. F. P. Nickless.
Report 73/15 ISBN 0 11 880658 0 £1.85

10 The sand and gravel resources of the country west of Colchester, Essex: Resource sheet TL 92. J. D. Ambrose.
Report 74/6 ISBN 0 11 880671 8 £1.45

11 The sand and gravel resources of the country around Tattingstone, Suffolk: Resource sheet TM 13. S. E. Hollyer.
Report 74/9 ISBN 0 11 880675 0 £1.95

12 The sand and gravel resources of the country around Gerrards Cross, Buckinghamshire: Resource sheet SU 99, TQ 08 and TQ 09. H. C. Squirrell.
Report 74/14 ISBN 0 11 880710 2 £2.20

Mineral Assessment Reports

13 The sand and gravel resources of the country east of Chelmsford, Essex: Resource sheet TL 70. M. R. Clarke.
ISBN 0 11 880744 7 £3.50

14 The sand and gravel resources of the country east of Colchester, Essex: Resource sheet TM 02. J. D. Ambrose.
ISBN 0 11 880745 5 £3.25

15 The sand and gravel resources of the country around Newton on Trent, Lincolnshire: Resource sheet SK 87. D. Price.
ISBN 0 11 880746 3 £3.00

16 The sand and gravel resources of the country around Braintree, Essex: Resource sheet TL 72. M. R. Clarke.
ISBN 0 11 880747 1 £3.50

17 The sand and gravel resources of the country around Besthorpe, Nottinghamshire: Resource sheet SK 86 and part of SK 76. J. R. Gozzard.
ISBN 0 11 880748 X £3.00

18 The Sand and gravel resources of the Thames Valley, the country around Cricklade, Wiltshire: Resource sheet SU 09/19 and parts of SP 00/10. P. R. Robson.
ISBN 0 11 880749 8 £3.00

19 The sand and gravel resources of the country south of Gainsborough, Lincolnshire: Resource sheet SK 88 and part of SK 78. J. H. Lovell.
ISBN 0 11 880750 1 £2.50

20 The sand and gravel resources of the country east of Newark upon Trent, Nottinghamshire: Resource sheet SK 85. J. R. Gozzard.
ISBN 0 11 880751 X £2.75

21 The sand and gravel resources of the Thames and Kennet Valleys, the country around Pangbourne, Berkshire: Resource sheet SU 67. H. C. Squirrell.
ISBN 0 11 880752 8 £3.25

22 The sand and gravel resources of the country north-west of Scunthorpe, Humberside: Resource sheet SE 81. J. W. C. James.
ISBN 0 11 880753 6 £3.00

23 The sand and gravel resources of the Thames Valley, The country between Lechlade and Standlake: Resource sheet SP 30 and parts of SP 20, SU 29 and SU 39. P. Robson.
ISBN 0 11 881252 1 £7.25

24 The sand and gravel resources of the country around Aldermaston, Berkshire: Resource sheet SU 56 and SU 66. H. C. Squirrell.
ISBN 0 11 881253 X £5.00

25 The celestite resources of the area north-east of Bristol: Resource sheet ST 68 and parts of ST 59, 69, 79, 58, 68 and 77. E. F. P. Nickless, S. J. Booth and P. N. Mosley.
ISBN 0 11 881262 9 £5.00

26 The limestone and dolomite resources of the country around Monyash, Derbyshire: Resource sheet SK 16. F. C. Cox and D. McC. Bridge.
ISBN 0 11 881263 7 £7.00

27 The sand and gravel resources of the country west and south of Lincoln, Lincolnshire: Resource sheets SK 95, SK 96 and SK 97. I. Jackson.
ISBN 0 11 884003 7 £6.00

28 The sand and gravel resources of the country around Eynsham, Oxfordshire: Resource sheet SP 40 and part of SP 41. W. J. R. Harries.
ISBN 0 11 884012 6 £3.00

29 The sand and gravel resources of the country south-west of Scunthorpe, Humberside: Resource sheet SE 80. J. H. Lovell.
ISBN 0 11 884013 4 £3.50

30 Procedure for the assessment of limestone resources. F. C. Cox, D. McC. Bridge and J. H. Hull.
ISBN 0 11 884030 4 £1.25

31 The sand and gravel resources of the country west of Newark upon Trent, Nottinghamshire: Resource sheet SK 75. D. Price and P. J. Rogers.
ISBN 0 11 884031 2 £3.50

32 The sand and gravel resources of the country around Sonning and Henley: Resource sheet SU 77 and SU 78. H. C. Squirrell.
ISBN 0 11 884032 0 £5.25

33 The sand and gravel resources of the country north of Gainsborough: Resource sheet SK 89. J. R. Gozzard and D. Price
ISBN 0 11 884033 9 £4.50

34 The sand and gravel resources of the Dengie Peninsula, Essex: Resource sheet TL 90, etc. M. B. Simmons.
ISBN 0 11 884081 9 £5.00

35 The sand and gravel resources of the country around Darvel, Strathclyde: Resource sheet NS 53, 63, etc. E. F. P. Nickless, A. M. Aitken and A. A. McMillan.
ISBN 0 11 884082 7 £7.00

36 The sand and gravel resources of the country around Southend-on-Sea, Essex: Resource sheets TQ 78, 79 etc. S. E. Hollyer and M. B. Simmons.
ISBN 0 11 884083 5 £7.50

37 The sand and gravel resources of the country around Bawtry, South Yorkshire: Resource sheet SK 69.
A. R. Clayton
ISBN 0 11 884053 3 £5.75

38 The sand and gravel resources of the country around Abingdon, Oxfordshire: Resource sheet SU 49, 59, SP 40, 50. C. E. Corser.
ISBN 0 11 884084 5 £5.50

39 The sand and gravel resources of the Blackwater Valley (Aldershot) area: Resource sheet SU 85, 86, parts SU 84, 94, 95, 96. M. R. Clarke, A. J. Dixon and M. Kubala.
ISBN 0 11 884085 1 £7.00

40 The sand and gravel resources of the country west of Darlington, County Durham: Resource sheet NZ 11, 21.
A. Smith.
ISBN 0 11 884086 X £5.00

41 The sand and gravel resources of the country around Garmouth, Grampian Region: Resource sheet NJ 36.
A. M. Aitken, J. W. Merritt and A. J. Shaw.
ISBN 0 11 884090 8 £8.75

42 The sand and gravel resources of the country around Maidenhead and Marlow: Resource sheet SU 88, parts SU 87, 97, 98. P. N. Dunkley.
ISBN 0 11 884091 6 £5.00

43 The sand and gravel resources of the country around Misterton, Nottinghamshire: Resource sheet SK 79.
D. Thomas and D. Price.
ISBN 0 11 884092 4 £5.25

44 The sand and gravel resources of the country around Sedgefield, Durham: Resource sheet NZ 32.
M. D. A. Samuel.
ISBN 0 11 884093 2 £5.75

45 The sand and gravel resources of the country around Brampton, Cumbria: Resource sheet NY 55, part 56. I. Jackson.
ISBN 0 11 884094 0 £6.75

46 The sand and gravel resources of the country around Harlow, Essex: Resource sheet TL 41. P. M. Hopson.
ISBN 0 11 884107 6 £9.50

47 The limestone and dolomite resources of the country around Wirksworth, Derbyshire: Resource sheet SK 25, part 35. F. C. Cox and D. J. Harrison.
ISBN 0 11 884108 4 £15.00

48 The sand and gravel resources of the Loddon Valley area: Resource sheet SU 75, 76, parts 64, 65, 66 and 74.
M. R. Clarke, E. J. Raynor and R. A. Sobey.
ISBN 0 11 884109 2 £8.75

49 The sand and gravel resources of the country around Lanark, Strathclyde Region: Resource sheet NS 94, part 84.
J. L. Laxton and E. F. P. Nickless.
ISBN 0 11 884112 2 £11.00

50 The sand and gravel resources of the country around Fordingbridge, Hampshire: Resource sheet SU 11 and parts of SU 00, 01, 10, 20 and 21. M. Kubala.
ISBN 0 11 884111 4 £7.75

51 The sand and gravel resources of the country north of Bournemouth, Dorset: Resource sheet SU 00, 10, 20, SZ 09, 19 and 29. M. R. Clarke.
ISBN 0 11 884110 6 £9.75

52 The sand and gravel resources of the country between Hatfield Heath and Great Waltham, Essex: Resource sheet TL 51 and 61. R. J. Marks.
ISBN 0 11 884113 0 £8.00

53 The sand and gravel resources of the country around Cottenham, Cambridgeshire: Resource sheet TL 46 and 47. A. J. Dixon.
ISBN 0 11 884114 9 £9.25

54 The sand and gravel resources of the country around Huntingdon and St Ives. Cambridgeshire: Resource sheets TL 16, 17, 26, 27, 36 and 37. R. W. Gatliff.
ISBN 0 11 884115 7 £8.75

55 The sand and gravel resources of the country around Ipswich, Suffolk: Resource sheet TM 14. R. Allender and S. E. Hollyer.
ISBN 0 11 884116 5 £10.00

56 Procedure for the assessment of the conglomerate resources of the Sherwood Sandstone Group. D. P. Piper and P. J. Rodgers.
ISBN 0 11 884143 2 £1.25

57 The conglomerate resources of the Sherwood Sandstone Group of the country around Cheadle, Staffordshire: Resource sheet SK 04. P. J. Rogers, D. P. Piper and T. J. Charsley.
ISBN 0 11 884144 0 *not yet priced*

58 The sand and gravel resources of the country west of Peterhead, Grampian Region: Resource sheet NK 04 and parts of NJ 94 and 95, NK 05, 14 and 15. A. A. McMillan and A. M. Aitken.
ISBN 0 11 884145 9 £12.00

59 The sand and gravel resources of the country around Newbury, Berkshire: Resource sheet SU 46 and 57, parts of SU 36, 37 and 47. J. R. Gozzard.
ISBN 0 11 884146 7 *not yet priced*

60 The sand and gravel resources of the country south-west of Peterborough, in Cambridgeshire and east Northamptonshire: Resource sheet TL 09 and 19 and SP 98 and TL 08. A. M. Harrisson.
ISBN 0 11 884147 5 £15.50

61 The sand and gravel resources of the country north of Wrexham, Clwyd: Resource sheet SJ 35 and part of SJ 25. P. N. Dunkley.
ISBN 0 11 884148 3 £11.75

62 The sand and gravel resources of the country around Dolphinton, Strathclyde Region, and West Linton, Borders Region: Resource sheet NT 04 and 14, parts of NT 05 and 15. A. A. McMillan, J. L. Laxton and A. J. Shaw.
ISBN 0 11 884149 1 £8.00

63 The sand and gravel resources of the valley of the Douglas Water, Strathclyde: Resource sheet NS 83 and parts of NS 82, 92 and 93. A. J. Shaw and E. F. P. Nickless.
ISBN 0 11 884150 5 £11.50

64 The sand and gravel resources of the country between Wallingford and Goring, Oxfordshire: Resource sheet SU 68 and part of SU 58. C. E. Corser.
ISBN 0 11 884151 3 *not yet priced*

65 The sand and gravel resources of the country around Hexham, Northumberland: Resource sheet NY 86 and 96. J. H. Lovell.
ISBN 0 11 884152 1 £7.50

66 The sand and gravel resources of the country west of Chelmsford, Essex: Resource sheet TL 60. P. M. Hopson.
ISBN 0 11 884153 X £8.50

67 The sand and gravel resources of the country around Hatfield and Cheshunt, Hertfordshire: Resource sheet TL 20 and 30, and parts of TQ 29 and 39. J. R. Gozzard.
ISBN 0 11 884167 X £10.00

68 The sand and gravel resources of the country north-east of Halstead, Essex: Resource sheet TL 83. R. J. Marks and J. W. Merritt.
ISBN 0 11 884168 8 £13.25

69 The sand and gravel resources of the country around Welwyn Garden City. Hertfordshire: Resource sheet TL 11 and 21. J. R. Gozzard.
ISBN 0 11 884169 6 £10.50

70 The sand and gravel resources of the country east of Harrogate, North Yorkshire: Resource sheet SE 35. D. L. Dundas.
ISBN 0 11 884170 7 £9.50

71 The sand and gravel resources of the country around Hemel Hempstead, St Albans and Watford: Resource sheet TL 00 and 10, and parts of TQ 09 and 19. W. J. R. Harries, S. E. Hollyer and P. M. Hopson.
ISBN 0 11 884171 8 *not yet priced*

72 The sand and gravel resources of the country around Bury St Edmunds, Suffolk: Resource sheet TL 86.
M. P. Hawkins
ISBN 0 11 884172 6 £10.50

73 The sand and gravel resources of the country between Ely and Cambridge, Cambridgeshire: Resource sheet TL 56, 57. A. R. Clayton.
ISBN 0 11 884173 4 £9.50

74 The sand and gravel resources of the country around Blaydon, Tyne and Wear: Resource sheet NZ 06, 16.
J. R. A. Giles.
ISBN 0 11 884174 2 *not yet priced*

75 The sand and gravel resources of the country around Stokesley, North Yorkshire: Resource sheet NZ 40, 50 and parts 41, 51. R. G. Crofts.
ISBN 0 11 884175 0 *not yet priced*

76 The sand and gravel resources of the country around Ellon, Grampian Region: Resource sheets NJ 93 with parts of NJ 82, 83 and 92, and NK 03 with parts of NK 02, and 13. J. W. Merritt.
ISBN 0 11 884176 9 £15.00

77 The limestone and dolomite resources of the country around Buxton, Derbyshire: Resource sheet SK 07 and parts of SK 06 and 08. D. J. Harrison.
ISBN 0 11 884177 7 £13.00

78 The sand and gravel resources of the country west of Boroughbridge, North Yorkshire: Resource sheet SE 36.
D. A. Abraham.
ISBN 0 11 884178 5 *not yet priced*

79 The limestone and dolomite resource of the country around Bakewell, Derbyshire: Resource sheet SK 26 and part of SK 27. D. McC. Bridge and J. R. Gozzard.
ISBN 0 11 884179 3 *not yet priced*

80 The sand and gravel resources of the country between Stamford and Peterborough: Resource sheet TF 00 and 10.
S. J. Booth.
ISBN 0 11 884180 7 £14.50

Reports of the Institute of Geological Sciences

Other Reports

69/9 Sand and gravel resources of the inner Moray Firth.
A. L. Harrison and J. D. Peacock.
ISBN 0 11 880106 6 35p

70/4 Sands and gravels of the southern counties of Scotland.
G. A. Goodlet.
ISBN 0 11 880105 8 90p

72/8 The use and resources of moulding sand in Northern Ireland. R. A. Old.
ISBN 0 11 881594 0 30p

73/9 The superficial deposits of the Firth of Clyde and its sea lochs. C. E. Deegan, R. Kirby, I. Rae and R. Floyd.
ISBN 0 11 880617 3 95p

77/1 Sources of aggregate in Northern Ireland (2nd edition).
I. B. Cameron.
ISBN 0 11 881279 3 70p

77/2 Sand and gravel resources of the Grampian Region.
J. D. Peacock and others.
ISBN 0 11 881282 3 80p

77/5 Sand and gravel resources of the Fife Region.
M. A. E. Browne.
ISBN 0 11 884004 5 60p

77/6 Sand and gravel resources of the Tayside Region.
I. B. Paterson.
ISBN 0 11 884008 8 £1.40

77/8 Sand and gravel resources of the Strathclyde Region.
I. B. Cameron and others.
ISBN 0 11 884028 2 £2.50

77/9 Sand and gravel resources of the Central Region, Scotland. M. A. E. Browne.
ISBN 0 11 884016 9 £1.35

77/19 Sand and gravel resources of the Borders Region, Scotland. A. D. McAdam.
ISBN 0 11 884025 8 £1.00

77/22 Sand and gravel resources of the Dumfries and Galloway Region of Scotland. I. B. Cameron.
ISBN 0 11 884021 5 £1.20

78/1 Sand and gravels of the Lothian Region of Scotland.
A. D. McAdam.
ISBN 0 11 884042 8 £1.00

78/8 Sand and gravel resources of the Highland Region.
W. Mykura, D. L. Ross and F. May.
ISBN 0 11 884050 9 £3.00

Dd 696431 K8

Typeset for the Institute of Geological Sciences by Bell and Bain Ltd., Glasgow.

Printed in England for Her Majesty's Stationery Office by Commercial Colour Press, London E7.